❧ The Poetry Friday Anthology for Science ❧

To Nancy —
Hooray for poetry
+ science
+ you!

# Additional Poetry Anthologies
## compiled by
## Vardell and Wong

### The Poetry Friday Anthology for Science
### Student Editions

The Poetry Friday Anthology for Science - Kindergarten
The Poetry Friday Anthology for Science - First Grade
The Poetry Friday Anthology for Science - Second Grade
The Poetry Friday Anthology for Science - Third Grade
The Poetry Friday Anthology for Science - Fourth Grade
The Poetry Friday Anthology for Science - Fifth Grade

### The Poetry Friday Anthology
### Grades K–5

Poems for the School Year
with Connections to the Common Core
or
with Connections to the TEKS

### The Poetry Friday Anthology for Middle School
### Grades 6-8

Poems for the School Year
with Connections to the Common Core
or
with Connections to the TEKS

### The PoetryTagTime Ebook Trio
PoetryTagTime (for K-5)
P*TAG (for teens)
Gift Tag (holiday poems)
the first-ever electronic anthologies
of original poetry
for children and teens

# THE POETRY FRIDAY ANTHOLOGY FOR SCIENCE

Poems for the School Year
Integrating Science, Reading, and Language Arts
K-5 Edition

218 poems by 78 poets

compiled by
꒜ Sylvia Vardell and Janet Wong ꒜

pomelo ✳ books

this book is dedicated to
all the poets
in our anthologies
for their generosity and inspiration

No part of this publication may be reproduced, or stored in a retrieval system, or transmitted in any form or by any means, electronic, mechanical, photocopying, recording, or otherwise, without written permission of the publisher. For information regarding permission, please contact us.

Pomelo Books
4580 Province Line Road
Princeton, NJ 08540
PomeloBooks.com
info@PomeloBooks.com

Copyright ©2014 by Sylvia M. Vardell and Janet S. Wong.
All rights reserved.

Individual poems copyright ©2014 by the individual poets; refer to credits. All rights reserved.

Library of Congress Cataloging-in-Publication Data is available.

ISBN 978-1-937057-97-8

Please visit us:
PomeloBooks.com

## Poems By

| | |
|---|---|
| Joy Acey | Renée M. LaTulippe |
| Alma Flor Ada | Debbie Levy |
| Linda Ashman | J. Patrick Lewis |
| Jeannine Atkins | George Ella Lyon |
| Carmen T. Bernier-Grand | Guadalupe Garcia McCall |
| Robyn Hood Black | Heidi Mordhorst |
| Susan Blackaby | Marilyn Nelson |
| Susan Taylor Brown | Kenn Nesbitt |
| Joseph Bruchac | Lesléa Newman |
| Leslie Bulion | Eric Ode |
| Stephanie Calmenson | Linda Sue Park |
| F. Isabel Campoy | Ann Whitford Paul |
| James Carter | Greg Pincus |
| Kate Coombs | Mary Quattlebaum |
| Cynthia Cotten | Heidi Bee Roemer |
| Kristy Dempsey | Michael J. Rosen |
| Graham Denton | Deborah Ruddell |
| Rebecca Kai Dotlich | Laura Purdie Salas |
| Shirley Smith Duke | Michael Salinger |
| Margarita Engle | Glenn Schroeder |
| Douglas Florian | Joyce Sidman |
| Betsy Franco | Buffy Silverman |
| Carole Gerber | Marilyn Singer |
| Charles Ghigna | Ken Slesarik |
| Joan Bransfield Graham | Eileen Spinelli |
| Mary Lee Hahn | Anastasia Suen |
| Avis Harley | Susan Marie Swanson |
| David L. Harrison | Carmen Tafolla |
| Terry Webb Harshman | Holly Thompson |
| Juanita Havill | Amy Ludwig VanDerwater |
| Esther Hershenhorn | Lee Wardlaw |
| Mary Ann Hoberman | Charles Waters |
| Sara Holbrook | April Halprin Wayland |
| Patricia Hubbell | Carole Boston Weatherford |
| Jacqueline Jules | Steven Withrow |
| Bobbi Katz | Allan Wolf |
| X. J. Kennedy | Virginia Euwer Wolff |
| Julie Larios | Janet Wong |
| Irene Latham | Jane Yolen |

"Science is the key to our future."

⁓ Bill Nye ⁓

# Table of Contents

| | |
|---|---|
| What Is Poetry Friday? | 8 |
| Poetry and Science | 11 |
| Where Does Poetry Fit? | 13 |
| The Science Curriculum and Science Standards | 15 |
| Reading Poetry Aloud | 16 |
| How to Use the *Take 5!* Box | 17 |
| Lexiles, Levels, and Standards | 18 |
| NGSS Disciplinary Core Ideas and Weekly Poems | 19 |
| A Note about Our Bilingual Poems | 20 |
| How to Use the Poems in this Book | 22 |
| A Poem for Everyone: "Scientific Steps" by Cynthia Cotten | 23 |
| Poems for Kindergarten | 25 |
| Poems for First Grade | 65 |
| Poems for Second Grade | 105 |
| Poems for Third Grade | 145 |
| Poems for Fourth Grade | 185 |
| Poems for Fifth Grade | 225 |
| A Poem for Everyone: "Science" by James Carter | 265 |
| Three Fun and Easy Ways to Celebrate Science Poetry | 266 |
| Building Your Own Poetry Library | 269 |
| Poetry Books for Science | 270 |
| Children's Poetry Websites and Blogs | 271 |
| Fun Websites to Support Science Learning | 273 |
| Websites and Blogs for Science Teaching (K-5) | 274 |
| Professional Resources | 275 |
| E-Resources for Poetry Teaching | 276 |
| A Mini-Glossary of Science Terms | 277 |
| Title Index | 282 |
| Poet Index | 285 |
| Subject Index | 287 |
| Copyright & Permissions | 291 |
| Poem Credits | 292 |
| About the Poets | 300 |
| TGIF! | 302 |
| Acknowledgments | 303 |
| About Sylvia Vardell and Janet Wong | 304 |
| A Note about Our Student Editions | 306 |

## What Is Poetry Friday?

In 2006 blogger Kelly Herold brought Poetry Friday to the "kidlitosphere." Much like "casual Friday" in the corporate world, there is a perception in the world of literature that on Fridays we should relax a bit and take a moment for something special. Why not bring the Poetry Friday concept into your classroom and take five minutes every Friday to share a poem and explore it a bit, connecting it with children's lives and capitalizing on a teachable moment? **In this book, we offer poetry that focuses particularly on science-themed topics.** Of course a poem alone is not intended to be an entire science lesson, but it offers a unique and innovative approach for beginning, complementing, or concluding a science lesson and an opportunity to build both science literacy and language learning in a fun and natural way.

On Poetry Friday you can find blog posts that include original poems, book reviews, poetry curriculum tips, and more. Each Friday a different blogger volunteers to gather and host a list of poetry posts from participating blogs. The list of who is hosting Poetry Friday each week can be found at A *Year of Reading* (ReadingYear.Blogspot.com) where two teachers and writers, Franki Sibberson and Mary Lee Hahn, post regularly about books, reading, and teaching.

Yes, of course you can share poetry on other days of the week, too—and we hope that you will whenever you want to highlight science topics or connect with a science lesson. But for those who are not already teaching poetry regularly, planning for Poetry Friday makes poetry sharing intentional and not incidental, with an added science bonus. Pausing to share a poem—and reinforce science learning—on Poetry Friday is an easy way to infuse poetry into your current teaching practice and to reinforce science content in increments. And once you have celebrated poetry and Poetry Fridays, we promise that students will be clamoring for it.

# Poetry and Science

"Everything must be made as simple as possible."

≈≈ Albert Einstein ≈≈

# Poetry and Science

This book is first and foremost a quality anthology of original poetry for children written by 78 of today's most popular poets. Children in any grade can enjoy, explore, and respond to these poems. However, we have also come to realize that educators, librarians, and parents are looking for guidance in how to share poetry with children and connect with the science curriculum at the same time. Thus, this book offers both: quality poetry organized by grade level plus curriculum-based suggestions for helping children enjoy and understand poetry AND science.

Like science, poetry often involves a high level of abstraction in language and ideas, requiring specific critical thinking skills and promoting interaction. In their 1994 *Reading Teacher* article, "Reciprocal processes in science and literacy learning," Casteel and Isom (1994) acknowledge, "The literacy processes are the means by which science content is learned because content information is rooted in written and oral language" (p. 540). Infusing poetry into the science curriculum can serve to jump-start or introduce a topic, present examples of terminology or concepts, provide closure that is concept-rich, or extend a science topic further. The brevity of poetry is less intimidating to children who may be overwhelmed by longer prose and streams of new vocabulary, especially students acquiring English as a new language. We can introduce or reinforce a science topic with a poem in just a few minutes with language that is rich, vivid, and memorable and activities that are engaging and interactive.

## Why Poetry with Science?

*Scientists observe with a clear eye, record their observations in precise, descriptive language, and craft their expressions. Poets do the same thing* (Bernice Cullinan, 1995). When it comes to supporting science instruction, poetry offers these distinct advantages:

- Poetry is accessible to a **wide range of reading abilities**.
- The brief format of much poetry **taps the essence of a subject.**
- Poetry incorporates **sensory language**, giving children the sense of touching, tasting, smelling, hearing, and seeing.
- Poetry can make a topic memorable through the use of **vivid imagery.**
- Poetry can provide a vehicle for content presented through evocative language and **rich vocabulary.**
- Poetry can help children **talk about issues** that concern them.
- **Science and literacy instruction can merge** through a one-minute poem.

Naturally, a single poem is not intended to be the entire science lesson, but it offers an innovative, engaging, vocabulary-full, and concept-rich way to launch or conclude a science lesson. Science expert Jill Castek challenges us to "break down those instructional silos" of science and literacy and look for opportunities to maximize overlap. We need to ensure that vocabulary exposure is occurring in many contexts for maximum scaffolding and science learning. In her essay, "Teaching science when your principal says 'teach language arts,'" Valarie Akerson notes, "It is possible to use language arts to support science learning and to use science as a purpose for learning language arts" (2002, p. 22). And Royce, Morgan, and Ansberry (2012) confirm "studies have shown gains in literacy as well as science achievement in programs that blend science and literacy instruction" (p. 6).

How can we share science poetry with young people while incorporating the *Next Generation Science Standards* (NGSS) and still maintain the joy and pleasure of poetry?

In *The Poetry Friday Anthology* series, we have borrowed the phrase "Take 5" from the great jazz musician Dave Brubeck to advocate taking time for poetry every Friday to introduce and share a poem—in this case a science-centric poem. Once again, **we provide "Take 5" activities for each poem to help you share the poem and invite students to listen and read with you, along with questions, activities, and book suggestions for considering the science content of each poem.**

The "Take 5" approach is based on a constructivist model of learning and encourages engagement and exploration in particular. The *Take 5!* activities provided are tied to the NGSS while also incorporating the literacy skills identified in the *Common Core State Standards* (CCSS). Obviously, poetry sharing doesn't take the place of planned science instruction, but the two complement each other well. You can also match poems and science lessons using the weekly themes or the index at the back of the book to identify relevant science topics.

As we consider the framework for the NGSS, we can explore how poetry might work alongside other texts and experiences to help students understand our "technology-rich and scientifically complex world." Akerson (2002) reminds us that the "use of language arts to promote literacy and support learning in other content areas is (also) recommended and encouraged by the International Reading Association (IRA) and the National Council of Teachers of English (NCTE)" (p. 22).

The more connections we can provide between what children are learning in various areas of study, the deeper their learning will be. If poetry can be that vehicle for connecting books, skills, concepts, and information across the curriculum, we owe it to children to infuse poetry wherever we can. We can encourage children to think like a poet AND a scientist in observing the world around them, using all their senses, seeing how things work, and gathering "big words" as they read, write, and learn.

# Where Does Poetry Fit?

You can share the poems in this book using any of the following approaches:

- Read aloud a poem once a week on Poetry Friday using the *Take 5!* activities to extend the science learning (in homeroom, reading class, science class, or in the library).

- Link the poem with science instruction directly by matching the poem's topic with the focus of the science lesson (using weekly themes or the subject index to select appropriate poems).

- Share a poem during reading or language arts instruction to emphasize the words, images, and language of the poem.

We can add poetry sharing to a planned science lesson by taking one minute to read aloud a science poem to set the stage for the instruction to come. Or conversely, ending with a poem can help reinforce the concepts introduced in a science lesson by building knowledge retention so crucial to learning. "Learning science and language is reciprocal" according to Casteel and Isom (1994), as children absorb new ideas and new vocabulary.

Using poetry in science can show children how writers approach scientific topics in very different and distinctive ways by starting a lesson with a science-themed poem. In addition, children will see that they can learn a lot of information from a poem. Poetry has an advantage in that it typically consists of fewer words than expository prose passages. Poems can be read and reread in very little time. Each rereading can be approached in a slightly different way—for example, through choral reading or poetry performance—and offer closure to a lesson or extend it further.

---

### Three Keys to Connecting Poetry and Science

In a variety of meaningful and participatory ways, we can celebrate poetry while gently introducing and reinforcing science knowledge across the grades. The keys to remember are:

- A poem should first be enjoyed for its own sake.

- Presenting poems in participatory ways (with various strategies) gets your learner "into the poem" and introduces the science content.

- The main idea is to help your learner think about science through the lens of a poem after enjoying the poem through multiple readings—and to increase exposure to science content as well as poetic language.

**Weekly Themes in *The Poetry Friday Anthology for Science***

The *Next Generation Science Standards* and the science standards of some states deliberately focus on a narrow selection of topics, favoring depth over breadth. With these poems, we hope to give students a glimpse of a wider selection of topics, to spur their curiosity, and to entice them to do supplemental science reading.

We offer 36 weeks of original poems for each grade level on the following topics. **These designated weekly themes cross all levels, K–5.** This provides a school-wide connection as each grade enjoys a different poem on the same topics.

- scientific practices
- lab safety
- questioning
- observations
- predictions & hypotheses
- investigations
- scientific tools
- data
- matter
- force, motion & energy
- light & sound
- space
- sun, earth & moon
- the water cycle
- weather & climate
- forces of nature
- soil & land
- natural resources
- ecosystems
- adaptations & traits
- cycles
- patterns
- the human body
- kitchen science
- video technology
- machines
- building things
- the science fair
- famous scientists
- science careers
- future challenges
- future dreams

## Poetry Breaks

Whether you introduce a poem at the beginning of the day, when transitioning to lunch or at a break, tied to a science lesson, or for wrapping things up, "breaking" for poetry provides a moment to refresh and engage. This doesn't mean that a more in-depth study of poetry as well as science is not a good idea. Of course it is. But for the average teacher or librarian, consistently sharing a five-minute poem break is an effective practice for injecting poetry into the routine. And with these science-themed poems, we offer the added bonus of infusing science content into this language experience. Communicate to students that a poetry break is about to begin by using a sign, bell, signal or chime announcing "Poetry Break!"

# The Science Curriculum and Science Standards

After several years of study, a multi-state consortium has developed the *Next Generation Science Standards* (NGSS), a detailed description of the key scientific ideas and practices that all students should learn by the time they graduate from high school.

These new standards are based on the National Research Council's *A Framework for K-12 Science Education*. The National Research Council, the National Science Teachers Association, the American Association for the Advancement of Science, and Achieve all partnered to create standards that are rich in content and practice.

*A Framework for K-12 Science Education: Practices, Crosscutting Concepts, and Core Ideas* principally concerns itself with "what all students should know in preparation for their individual lives and for their roles as citizens in this technology-rich and scientifically complex world."

This framework addresses "major challenges that confront society today, such as generating sufficient energy, preventing and treating diseases, maintaining supplies of clean water and food, and solving the problems of global environmental change."

There are three key dimensions of the framework—Practices, Concepts, and Core Ideas:

1. **Scientific and Engineering Practices** (e.g., asking questions, defining problems, using models, planning investigations, analyzing data, communicating information)

2. **Crosscutting Concepts** (e.g., patterns, cause and effect, scale, proportion and quantity, systems, energy and matter, structure and function, and stability and change)

3. **Disciplinary Core Ideas** (across the physical sciences, life sciences, earth and space sciences, engineering, technology, and applications of science)

We used this framework to guide us in organizing this anthology of poems, particularly in designating weekly science-based themes for the poems and for targeting science learning in the *Take 5!* activities provided for every poem.

We also provide a matrix showing how the poems in this book meet those practices and core ideas at each grade level. We should also note that different states and communities may rely on different state and local curriculum standards to guide science instruction. (For example, check out PoetryForScience.Blogspot.com for tables that correlate the poems in this book with the TEKS Science and Technology standards in Texas.) There is significant overlap across states and the poems and activities provided here offer meaty and meaningful guidance in either case.

# Reading Poetry Aloud

A guiding principle of this book is that poetry is meant to be read aloud. As the award-winning poet Eve Merriam noted, "It's easier to savor the flavor of the words as they roll around in your mouth for your ears to enjoy." As with song lyrics that sit quietly on the page, the music of poetry comes alive when spoken and shared. It is also the ideal way to approach poetry instruction that builds on students' comfort and familiarity with spoken language.

The more children hear, read, say, and experience the poem, the more they internalize the sounds, words, and meanings of the poem and can focus on its content.

### Ten Easy Tips for Reading Poetry Aloud

Here are some tips to help you read poetry aloud effectively (drawn from the work of Jack Collom and Sheryl Noethe in *Poetry Everywhere*):

1. Be sure to **say the title and author** of the poem.
2. If possible, **display the words** of the poem while you read it aloud.
3. **Be yourself.** You needn't and shouldn't show reverence for poetry by means of an artificially dignified atmosphere.
4. **Energy is the key**—but it shouldn't be forced. A brisk pace is good, as long as you slow down when the situation needs it.
5. Practice reading with **pauses at the ends of lines** and in other places; think about how this affects your students' understanding of a poem.
6. Be sure to **enunciate each word distinctly** and check uncertain pronunciations beforehand.
7. **Glance at the audience** occasionally.
8. It's helpful to **admit your own errors and ignorance**—about poetry and about science. Don't worry. Decide what you're going to do and go for it.
9. Relax and concentrate. Freely **intersperse humor and seriousness.** Have fun!
10. **Pause for a moment after the poem reading** to let the words and content sink in.

You can find many more tips for performing poetry in fun and easy ways in *The Poetry Teacher's Book of Lists* by Sylvia Vardell.

## How to Use the *Take 5!* Box

Each poem in this book faces a *Take 5!* box of teaching tips. Below you will find the *Take 5!* box for the poem "Capillary Action" by Joy Acey (Kindergarten, Week 6, page 34).

#1: Here you will find an easy suggestion for **how to make the poem come alive** as you read it aloud by pairing the poem with a prop, adding gestures or movement, trying specific dramatic reading techniques, adding multi-media, and so on.

#2: This tip suggests **how to engage students in participating with you** in reading the poem aloud again. For example, look for any repeated words, phrases, lines, or stanzas in the poem and invite students to chime in on those words as you read the rest of the poem aloud.

### Take 5!

1. Before or after reading this poem aloud, you may need to **explain the key concept of "capillary action."** Put a **drinking straw in a glass of water and watch some water climb up the straw to show how liquid can move through tiny, built-in "tubes"** like this. Then read the poem aloud again. Or show students the experiment at Education.com/science-fair/article/capillary-action/.

2. Read the poem aloud again and **invite students to chime in when the title of the poem appears within the poem** (*Capillary Action*). Cue them by cupping your hand behind your ear.

3. For discussion: *How can you tell when a plant is thirsty?*

4. **Show examples of capillary action (including the celery experiment) from the USGS Water Science School online** at GA.Water.USGS.gov/edu/capillaryaction.html.

5. Pair this with another poem about a simple investigation that involves plants, **"First Science Project" by Lesléa Newman** (1st Grade, Week 26, page 94), or look for Rebecca Kai Dotlich's book *What Is Science?*

#3: You'll find **a fun discussion prompt** here, tailored to fit the poem. It's usually an open-ended question with no single, correct answer. Encourage diversity in responses!

#4: Here we **connect the poem to a specific science skill or concept** offering a targeted focus for quick explanation, simple demonstration, or multimedia connection.

#5: We share **related poem titles and book titles** that connect well with the featured poem based on the poem content or science topic.

While the content for each *Take 5!* box is tailored to the individual poems, all *Take 5!* boxes contain the same sequence of teaching tips for consistency and ease of use. Again, these are only suggested activities and curriculum connections. You'll want to pick and choose which suggestions work best for your students.

## Lexiles, Levels, and Standards

The readability level of poems varies greatly since poetry doesn't easily fit the use of Lexiles and levels. Simple poems can have very sophisticated vocabulary and long poems can use simple language. Determining Lexile levels is based on a variety of factors such as how long the sentences are and how unusual the words are, as well as on the use of basic punctuation. The nursery rhyme "Little Jack Horner," for example, is written at the same eighth grade level as Robert Frost's classic poem "Stopping by Woods on a Snowy Evening." But clearly they are significantly different works. The power lies in a poem's meaning and in the distinctive ways the poet uses and arranges words. With these principles in mind, this anthology presents poems for each grade level selected for their relevance, interest, and appropriateness for each grade. In addition, the poems highlight the Next Generation Science Standards (NGSS) Disciplinary Core Ideas that are outlined for each grade level, kindergarten through fifth grade. The grid (opposite) indicates which of these core ideas is featured in the poem for each week at each grade level.

In addition, this exposure to weekly poetry naturally addresses the skills for reading and language learning. Many of the Common Core State Standards (CCSS), for example, are also met when we share these science-centric poems. The poems in this anthology contain rhyme, rhythm, sensory language, repetition, alliteration, imagery, similes, metaphors, and more—all elements that children need to encounter as they develop as readers. For more information on which poems incorporate which CCSS areas at each grade level, check our blog, PoetryForScience.Blogspot.com.

## Conclusion

Read these poems aloud so that students of all abilities and language learning levels can participate in them. Display the words so that students can have visual as well as aural reinforcement. Invite students to join in saying words and reading the text as they make the poem their own. Talk about the poems and invite students to share what they notice and build on their science knowledge.

Link with more poems, more poets, more poetry books, science-themed nonfiction, and relevant websites. Infuse poem-sharing throughout the day and throughout the science curriculum. Mark this book up with notes about your students' responses to individual poems and with science lesson connections. And don't be surprised if it's a wonderful poem moment that students remember most vividly at the end of the school year!

# Next Generation Science Standards (NGSS): Disciplinary Core Ideas and Weekly Poems

|  | **Processes, Scientists & Safety*** | **Life Science** | **Earth & Space Science** | **Physical Science** | **Engineering** |
|---|---|---|---|---|---|
| **K** | Discipline of Science; Lab Safety<br><br>Weeks 1, 2, 3, 4, 5, 6, 7, 8, 30, 31, 32 | Interdependent Relationships in Ecosystems: *Animals, Plants & Their Environment*<br><br>Weeks 4, 6, 21, 22, 23, 24, 25, 26, 31, 32, 34 | Weather, Climate & Changes<br><br><br><br>Weeks 4, 9, 14, 15, 16, 17, 18, 19, 35 | Forces & Interactions: *Pushes & Pulls*<br><br><br><br>Weeks 5, 10, 11, 12, 13 | K-2 Engineering Design<br><br><br><br>Weeks 20, 27, 28, 29, 33, 36 |
| **1st** | Discipline of Science; Lab Safety<br><br>Weeks 1, 2, 3, 5, 7, 30, 35 | Structure & Function<br><br>Weeks 4, 6, 16, 19, 20, 21, 22, 23, 25, 26, 35 | Space Systems: *Patterns & Cycles*<br><br>Weeks 14, 15, 16, 17, 18, 24, 36 | Waves: *Light & Sound*; Matter<br><br>Weeks 5, 8, 9, 10, 11, 13, 28, 34 | K-2 Engineering Design<br><br>Weeks 12, 19, 20, 27, 29, 31, 32, 33 |
| **2nd** | Discipline of Science; Lab Safety<br><br>Weeks 1, 2, 3, 4, 5, 6, 7, 8, 17, 30 | Interdependent Relationships in Ecosystems<br><br>Weeks 5, 19, 20, 21, 22, 23, 24, 25, 26, 31, 32, 35 | Earth's Systems: *Processes that Shape the Earth*<br><br>Weeks 12, 14, 15, 16, 17, 18, 19, 20 | Structure & Properties of Matter<br><br>Weeks 2, 9, 10, 11, 13, 28, 30 | K-2 Engineering Design<br><br>Weeks 6, 27, 28, 29, 33, 34, 36 |
| **3rd** | Discipline of Science; Lab Safety<br><br>Weeks 1, 2, 3, 4, 5, 11, 30, 31, 35 | Interdependent Relationships in Ecosystems; Inheritance & Variation of Traits: *Life Cycles & Traits*<br><br>Weeks 8, 16, 19, 21, 22, 23, 25 | Weather & Climate<br><br><br><br>Weeks 14, 15, 17, 18, 20, 31, 32 | Forces & Interactions; Light Waves<br><br><br><br>Weeks 6, 7, 9, 10, 11, 13, 20, 26, 29 | 3-5 Engineering Design<br><br><br><br>Weeks 7, 12, 27, 28, 32, 33, 34, 36 |
| **4th** | Discipline of Science; Lab Safety<br><br>Weeks 1, 2, 3, 4, 5, 7, 32, 33 | Structure & Function<br><br>Weeks 4, 6, 19, 21, 22, 23, 24, 25, 26, 30, 34, 35 | Earth's Systems: *Processes that Shape the Earth*<br><br>Weeks 14, 15, 16, 17, 18 | Energy & Waves: *Waves & Information*<br><br>Weeks 9, 10, 11, 12, 13, 20, 28, 31 | 3-5 Engineering Design<br><br>Weeks 8, 20, 27, 28, 29, 33, 36 |
| **5th** | Discipline of Science; Lab Safety<br><br>Weeks 1, 2, 3, 6, 7, 26, 30 | Matter & Energy in Organisms & Ecosystems<br><br>Weeks 4, 5, 8, 15, 16, 20, 21, 22, 23, 24, 25, 33, 34, 35 | Earth & Space Sys.: *Stars & Solar System*<br><br>Weeks 14, 15, 16, 17, 18, 19, 20, 32 | Structure and Properties of Matter<br><br>Weeks 4, 6, 9, 10, 11, 12, 13, 31 | 3-5 Engineering Design<br><br>Weeks 8, 27, 28, 29, 33, 36 |

*****Processes, Scientists & Safety.** This topic is not one of the "disciplinary core ideas" included in the NGSS, but we believe it is important in setting the stage for introducing science-themed poetry with children.

## A Note about Our Bilingual Poems

In this anthology, we have also included several bilingual poems (English/Spanish) by Alma Flor Ada, Carmen T. Bernier-Grand, F. Isabel Campoy, Margarita Engle, Guadalupe Garcia McCall, and Carmen Tafolla. These poems offer opportunities to share both poetry and science content with students in two languages. Keep in mind, however, that poems that rhyme in one language may not rhyme in the translated version; they can be parallel in meaning, but not literal word-for-word translations in order to maintain the musicality of the poem.

As you introduce students to these gems, you can read them aloud in either or both languages, depending on your own comfort and linguistic expertise. If you work with students, colleagues, or parents who are fluent Spanish speakers, this may be an opportunity to invite them to participate.

But be sure **not** to single out a particular child or staff member, and to ask in general for a volunteer. Don't assume they want to read aloud any more than a fluent English speaker wants to volunteer to read a poem aloud. Many Spanish-surnamed or Latino-origin children do not speak Spanish or do not feel comfortable being expected to read in Spanish. Set the stage by reading these poems aloud yourself and inviting students to read aloud with you, using the various strategies provided in the *Take 5!* activities.

As both students and teacher get comfortable with hearing and sharing poems, create an inviting atmosphere that encourages everyone to take a chance and raise their voices in poetry reading.

# The Poems

## How to Use the Poems in this Book

We offer a set of 36 poems (a poem-a-week for the nine months of the typical school year) for each grade level and have designed activities that are poem-specific, science-focused, and developmentally appropriate for each weekly poem. But there are several ways you could use the poems in this book.

- You can simply find your preferred grade level and share a poem-a-week from Week 1 to Week 36 any time as a bonus plug for science content.

- You can jump around the book, seeking out poems that fit with specific science lessons using the weekly themes (such as lab safety, weather and climate, etc.) or the subject index provided in the back of the book to locate the perfect poem for your lesson topic.

- You can share these poems (from your designated grade level or from anywhere in the book) during your usual Reading/Language Arts period, infusing an added science focus into your poem sharing.

Our goal is to provide support for educators and parents who might be unfamiliar with today's poetry for young people and might need guidance in how to begin while weaving science content throughout our poetry sharing. To keep the poetry momentum going, for each poem you share we suggest another poem from the book that is related in some way. Feel free to share any and all of the poems with the students you teach at any time, in any order, and in any way you like.

Note that there are two easy ways to share poems with students:

- Project an e-book version of our book from your computer onto a screen or Smartboard.

- Have students follow along in one of our Student Editions, with Bonus Poems for each grade. See pages 306-307 for more information about our Student Editions.

### SCIENTIFIC STEPS
by **Cynthia Cotten**

Find a problem, ask a question.
That's the way you start.
Now do research, all you can.
That's the second part.
Predict an answer—a hypothesis—
Based on what you know.
Run a test—an experiment.
Finished? How'd it go?
Repeat that step a few more times.
Are your results the same?
Analyze your data—
Does it back up your claim?
Write up your observations.
Be clear, avoid confusion.
Then share with others what you've found.
This is your conclusion.

"Now is the time to understand more, so that we may fear less."

⁂ Marie Curie ⁂

# Poems for Kindergarten

# NGSS Science and Engineering Practices: Kindergarten

*These practices form the foundation of disciplinary literacy in science and integrate reading, writing, listening, and speaking skills from the language arts. Here we indicate which weekly poems emphasize which science and engineering practices at each grade level.*

| PRACTICE | POEM |
| --- | --- |
| Asking questions and defining problems | Weeks 1, 3, 14, 20, 27, 35 |
| Developing and using models | Weeks 9, 28, 36 |
| Planning and carrying out investigations | Weeks 2, 5, 29 |
| Analyzing and interpreting data | Weeks 7, 17, 22, 24, 34 |
| Using mathematics and computational thinking | Week 8 |
| Constructing explanations and designing solutions | Weeks 6, 16, 19 |
| Engaging in argument from evidence | Weeks 10, 11, 23, 26, 30, 32 |
| Obtaining, evaluating, and communicating information | Weeks 4, 12, 13, 15, 18, 21, 25, 31, 33 |

# KINDERGARTEN

| | | |
|---|---|---|
| week 1 | Scientific Practices | When You Are a Scientist *by Eric Ode* |
| week 2 | Lab Safety | Superhero Scientist *by Joan Bransfield Graham* |
| week 3 | Ask and Ask Again | I Have a Question *by Anastasia Suen* |
| week 4 | Observations | Step Outside. What Do You See? *by Allan Wolf* |
| week 5 | Predictions & Hypotheses | Sink or Float *by Janet Wong* |
| week 6 | Investigations | Capillary Action *by Joy Acey* |
| week 7 | Data | My Bean Plant *by Amy Ludwig VanDerwater* |
| week 8 | Tools of Science | Stopwatch *by Janet Wong* |
| week 9 | Matter | Water + Dirt = *by Rebecca Kai Dotlich* |
| week 10 | More Matter | Take Backs *by Janet Wong* |
| week 11 | Force, Motion & Energy | Push Power *by Janet Wong* |
| week 12 | More FM&E | Thank You, Isaac Newton *by Eileen Spinelli* |
| week 13 | Light & Sound | Listen *by Amy Ludwig VanDerwater* |
| week 14 | Space | Did You Know? *by Julie Larios* |
| week 15 | Sun, Earth & Moon | Big Sun *by Douglas Florian* |
| week 16 | The Water Cycle | Old Water *by April Halprin Wayland* |
| week 17 | Weather & Climate | Dog in a Storm *by Stephanie Calmenson* |
| week 18 | Forces of Nature | Riddle for a Dry Day *by Irene Latham* |
| week 19 | Soil & Land | My Rock *by Ken Slesarik* |
| week 20 | Natural Resources | Auntie V's Hybrid Car *by Janet Wong* |
| week 21 | Ecosystems | Alligator with Fish *by Jane Yolen* |
| week 22 | Adaptations & Traits | Snake Traits *by Linda Ashman* |
| week 23 | Cycles | Young & Old Together/Jóvenes y viejos juntos *by Margarita Engle* |
| week 24 | Patterns | Inherit Tense *by Charles Ghigna* |
| week 25 | Human Body | Hands *by Kate Coombs* |
| week 26 | Kitchen Science | Can Our Eyes Fool Our Taste Buds? *by April Halprin Wayland* |
| week 27 | Video Technology | Hello, Hello! *by Janet Wong* |
| week 28 | Machines | Metal Monster *by X.J. Kennedy* |
| week 29 | Building Things | Tinker Time *by Janet Wong* |
| week 30 | Science Fair | Science Fair Day *by Eric Ode* |
| week 31 | Famous Scientists | Rachel Carson *by Julie Larios* |
| week 32 | More Famous Scientists | Ocean Explorer Sylvia Earle *by Leslie Bulion* |
| week 33 | Computers | Computer Geek/Compu-nerdo *by Carmen T. Bernier-Grand* |
| week 34 | Science Careers | Dr. Lee *by Janet Wong* |
| week 35 | Future Challenges | Water *by Kate Coombs* |
| week 36 | Future Dreams | Future Dreams Idea #63 *by Janet Wong* |

*"Men **love to wonder**, and that is the seed of science."*

∽ Ralph Waldo Emerson ∾

## Week 1: Scientific Practices

### Take 5!

1. **Draw a question mark on the board** and talk about how everybody likes to ask questions, particularly scientists and poets, who like to wonder about things. Then read this poem aloud, pausing briefly at the end of each line.

2. The last line of each stanza repeats a key word (*why, try*) three times. **Invite students to chime in on those repeated words (*why, try*)** as you read the whole poem aloud again. Pump your fist in the air as a cue for students to join in.

3. For discussion: **Which is more important: questions or answers?** Talk about how BOTH are important.

4. Invite students to **talk about what is involved in being a scientist based on the steps presented in this poem**: Asking questions (like what, when, how, where, and why) and finding answers (by reading, watching, thinking, writing, and trying repeatedly). Talk about how each of these is an important part of solving problems.

5. Share another poem about asking questions, **"Backwards" by Janet Wong** (1st Grade, Week 3, page 71), or connect with a simple introductory nonfiction book like *What Is a Scientist?* by Barbara Lehn or *You Are a Scientist* by Marcia S. Freeman.

## When You Are a Scientist
by **Eric Ode**

When you are
a scientist,
ask what
and when
and how
and where
and why, why, why.

When you are
a scientist,
read,
and watch,
and think,
and write,
and try, try, try.

# KINDERGARTEN

## WEEK 2: LAB SAFETY

### SUPERHERO SCIENTIST
by **Joan Bransfield Graham**

As a Superhero Scientist,
I need some special gear.

I wash my hands, put on gloves—
safety goggles always near,

right next to my microscope,
on a handy shelf.

If I want to save the world,
I have to save myself.

### Take 5!

1. Safety goggles are a very important part of scientific investigation, so if you have access to a pair, this is an ideal moment to demonstrate their use, **putting on safety goggles while you read the poem aloud.** If not, simply pantomime the actions in the poem (putting on gloves, reaching for goggles, putting on goggles).

2. This time, **invite students to "echo read" the last two lines of the poem while you read the rest aloud.** You begin by reading the first three couplets (first six lines) aloud, then pause and read the seventh line (*If I want to save the world*). Cup your ear to cue students to repeat the line, then read the final line (*I have to save myself*) and cue students to repeat that line, too.

3. For discussion: *What else can we do to stay safe in the classroom science lab?* (Hand washing, use of safety goggles, etc.)

4. **When it comes to science experiments and investigations, it's important to begin by talking about science safety procedures.** Research what your school, district, or community mandates for classroom and outdoor investigations such as wearing safety goggles, washing hands, following directions, and using materials and equipment appropriately, and make a list or poster highlighting those guidelines.

5. Connect this poem with another about lab safety, **"The Science Lab Pledge" by Deborah Ruddell** (1st Grade, Week 2, page 70).

## WEEK 3: ASK AND ASK AGAIN

### Take 5!

1. **Before reading this poem aloud, hold up a giant cut-out question mark** and ask students to guess what this poem is about. Then read the poem aloud.

2. **Help students memorize the repeated lines** *"a question"* and *"the answer"* and **invite them to chime in on those lines** while you read the rest of the poem aloud. Cue them with your giant cut out question mark.

3. **Talk about the concept of "opposite"**—like question/answer—and brainstorm other examples of opposites.

4. **Create a "Question Wall" by brainstorming a class list of science-focused questions** and posting them on craft paper covering a door or portion of a wall. As students visit the library or you choose books to read aloud, make connections with the questions they are raising.

5. Share another poem with many examples of questions, **"Late Night Science Questions" by Greg Pincus** (2nd Grade, Week 3, page 111).

### I Have a Question
by **Anastasia Suen**

I have a question,
a question,
a question.
I have a question.
I need to know why.

I look for the answer,
the answer,
the answer.
I look for the answer.
I give it a try.

## WEEK 4: OBSERVATIONS

**KINDERGARTEN**

### STEP OUTSIDE. WHAT DO YOU SEE?
by **Allan Wolf**

Step outside. What do you see?
A bird. A bug. A bumblebee.
A leaf. A breeze. A budding tree.

Step outside. The weather's fine.
A rock. A stick. A tall green pine.
Let's go out, rain or shine!

### Take 5!

1. **Bring something from the outdoors inside as a poetry prop to show while you read the poem aloud.** If possible, bring something cited in the poem like a leaf, rock, or stick.

2. In a repeated reading of the poem, **invite students to echo you after each PHRASE of the poem (rather than each LINE of the poem)** as you read the poem aloud. Pause after each period, comma, question mark or exclamation mark, such as after *Step outside*, *What do you see?*, *A bird*, *A bug*, *A bumblebee*, *A leaf*, *A breeze*, *A budding tree*, etc.

3. Work together to **create a quick poem collage, inviting students to choose their favorite images for each item cited in the poem, either drawings or magazine pictures** (bird, bug, bumblebee, leaf, breeze, budding tree, rock, stick, tall green pine, rain, sunshine). Talk about each item as you challenge students to put them in the order they are mentioned in the poem.

4. If there is a window nearby, **invite students to spend a minute observing the organisms, objects, and events they see outside.** Make a quick list of those items and ask students to group or classify the things they see in some way. Even better, take a moment to step outside for this activity, if possible.

5. Share a similar poem about exploring the natural world with **"Discovery / Descubrimiento" by Margarita Engle** (2nd Grade, Week 4, page 112), or look for selections from *Outside Your Window: A First Book of Nature* by Nicola Davies.

## Week 5: Predictions & Hypotheses

**Kindergarten**

### Take 5!

1. If possible, **gather a few objects mentioned in the poem (popsicle stick, crayon, straw, penny), as well as a cup or bowl of water, to show as your poetry props.** Then read the poem aloud.

2. As you share the poem another time, **invite the students to chime in on the opening phrase of the final line, *Try it,*** while you read the rest of the poem aloud.

3. Guide the students in guessing **which items in the poem are likely to float and why.** Consider other objects they have seen floating or sinking in tubs or pools, etc.

4. **If possible, try to conduct the simple investigation described in the poem.** One resource is Education.com/activity/article/sinkorfloat_kindergarten/. Make a group pictograph to share the results and show the data.

5. For another poem about objects that float, look for **"Everyday Astronaut / Un astronauta común" by Carmen Tafolla** (1st Grade, Week 36, page 104).

### Sink or Float
by **Janet Wong**

Popsicle stick.
Crayon. Straw.
Brand new shiny penny.

I'm guessing which
will float like a boat—
I wonder which will. Any?

Take a guess:
what do you think?
Try it—check if it will sink!

## Week 6: Investigations

### Capillary Action
by **Joy Acey**

I put my stick of celery
in my cherry drink.
Three days later
the leaves turned pink!

Tell me your reaction.
Tell me what you think.
Could capillary action
happen when I drink?

### Take 5!

1. Before or after reading this poem aloud, you may need to **explain the key concept of "capillary action."** Put a **drinking straw in a glass of water and watch some water climb** up the straw to **show how liquid can move through tiny, built-in "tubes"** like this. Then read the poem aloud again. Or show students the experiment at Education.com/science-fair/article/capillary-action/.

2. Read the poem aloud again and **invite students to chime in when the title of the poem appears within the poem** (*Capillary Action*). Cue them by cupping your hand behind your ear.

3. For discussion: *How can you tell when a plant is thirsty?*

4. **Show examples of capillary action (including the celery experiment) from the USGS Water Science School online** at GA.Water.USGS.gov/edu/capillaryaction.html.

5. Pair this with another poem about a simple investigation that involves plants, **"First Science Project" by Lesléa Newman** (1st Grade, Week 26, page 94), or look for Rebecca Kai Dotlich's book *What Is Science?*

## WEEK 7: DATA

### KINDERGARTEN

### MY BEAN PLANT
by **Amy Ludwig VanDerwater**

I made a graph so I can show
how every week my plant will grow.

It's planted in a paper cup
and every week my graph goes up.

I can tell you what this means.
I am good at growing beans.

### Take 5!

1. **Show a simple graph as a poetry prop** for this poem before reading the poem aloud.

2. In sharing this poem again, **read the poem aloud and pause before the final word, inviting students to finish it** with the word *beans*.

3. Work together to **create a quick, simple graph of student data**—such as those who like and don't like beans.

4. **It would be ideal to provide each student with a cup or empty milk carton, a bean, and soil and have them try this simple experiment themselves.** Or plant a single bean in a cup (or two or three beans in two or three cups) to monitor as a class. Then you can work together to record information about the plant's growth or lack of growth over a period of days or weeks through drawings, words, or measurements.

5. Follow up this poem with another about growing plants, **"Pumpkin Experiment" by Mary Lee Hahn** (1st Grade, Week 6, page 74), or look for selections from George Shannon's poetry book, *Busy in the Garden* or Gail Gibbons' nonfiction book, *From Seed to Plant*.

## WEEK 8: TOOLS OF SCIENCE

**KINDERGARTEN**

### STOPWATCH
by **Janet Wong**

We start timing
with a click—

then

tick tick tick tick tick tick tick tick tick tick
tick tick tick tick tick tick tick tick tick tick
tick tick tick tick tick tick tick tick tick tick

we click again.

Thirty seconds just passed?
Oh my!
Our stopwatch
counts *so* fast!

### Take 5!

1. **Ideally, have a stopwatch handy to use as your poetry prop** for this poem. Read the poem aloud, clicking the button when you read the lines, *"tick, tick tick..."* Or use the 30-second stopwatch recording from SoundSnap.com or a stopwatch app on your cell phone, if you have one.

2. Read the poem aloud again and **invite the students to say the *tick tick tick* lines with you.** (*Tick* occurs 30 times = 30 seconds.)

3. For discussion: **When is it handy to have a stopwatch?**

4. **Talk about how scientists use clocks, timers, and stopwatches** to study the natural world and collect information. Try using a clock or stopwatch for an on-the-spot investigation counting the seconds of uninterrupted quiet in the room, for example.

5. For another poem about using time to measure data, share **"Crazy Data Day" also by Janet Wong** (2nd Grade, Week 7, page 115). Also look for these poetry collections focused on the topic of time: *It's About Time* edited by Lee Bennett Hopkins and *First Morning: Poems About Time* edited by Nikki Siegen-Smith.

## WEEK 9: MATTER

### KINDERGARTEN

### WATER + DIRT =
by **Rebecca Kai Dotlich**

**Mud.**
I will make some for us.
Easy! No fuss.
A handful of dirt.
Some water—squirt!
Need one small stick.
Stir until thick.
What does mud make?
A puddle.
A cake.
And goodness knows,
it feels good between toes.

### Take 5!

1. Ask students if they know what a recipe is—a list of ingredients and step-by-step directions for how to make something. **Tell them to listen for the *ingredients* and *directions* in this poem.** Then read the poem aloud slowly and with expression.

2. **Coach students to answer the question raised in the poem (*What does mud make?*) by saying the lines *A puddle* and *A cake*.** Read the whole poem aloud again, pausing after the question line so the students can chime in. Then finish reading the poem aloud.

3. For discussion: *What are the good and bad things about mud?*

4. **Use this poem to talk about the different qualities of mud.** It starts as soil or dirt, you add a squirt of water, and then you stir it with a stick. When is it a puddle? When is it a "cake?" Talk about the amount of soil and water needed to make each. Describe the size, shape, color, and texture of a puddle vs. a "cake." Bring a cup of soil or dirt and a cup of water to demonstrate, if possible.

5. Pair this poem with another about mixing ingredients, **"Sugar Water" by Janet Wong** (1st Grade, Week 10, page 78), or look for selections from the classic book *Mud Pies and Other Recipes* by Marjorie Winslow.

## Week 10: More Matter

### Kindergarten

### Take Backs
by **Janet Wong**

Some things you can take back
without any trouble—
like sand mixed with rocks
or peas mixed with rice.

Some things you can't take back—
like salt in cake batter.
Grandpa thought
he grabbed the sugar but
he put a cup of salt
in by accident—

no taking that back, for sure—
especially after the batter's been baked!

Want a slice of Salty Chocolate Cake?

### Take 5!

1. If possible, **have a packet of salt and a packet of sugar handy as a poetry prop. Tear them open and pour out the contents.** Ask students if they know the difference between these two items before reading the poem aloud.

2. Read the poem aloud again and **invite students to chime in with an answer to the question that concludes the poem,** *Want a slice of Salty Chocolate Cake?* Talk about why they might answer no—or yes.

3. **Invite students to share stories of helping adults or older siblings or friends cook or bake,** particularly if any goofs or mistakes were made!

4. Talk about **how mixing things together can change both things, especially when you heat them or cool them.** Use the cake in this poem as an example. (Mixing salt in the cake batter causes the salt to blend with the other ingredients, but still maintain its salty flavor, even when baked.)

5. Follow up this poem with another poem about food and science, **"Breakfast Alchemy" by Mary Quattlebaum** (3rd Grade, Week 26, page 174), or bring in a simple cookbook to talk about how recipes involve both science and math.

## Week 11: Force, Motion & Energy

**Kindergarten**

### Push Power
by **Janet Wong**

I pull with my hands.
My wagon is stuck.
I push harder with legs.
This time I'm in luck.
My wagon gets out
of the mud
but—
wait!

It zooms
down the hill
straight into the lake!

### Take 5!

1. **Use pantomime to bring this poem to life,** acting out the pulling of the wagon with your hands, then digging in with your feet and pulling again, then with fists cheering, then with hands gesturing *wait*, then running after the runaway wagon.

2. In a repeated reading of this poem, **coach students to say or even shout the word *wait* when it occurs in the poem.** You read the rest of the poem aloud, pausing after the word *but* and signaling students with both palms out, motioning *wait*, as their cue to chime in.

3. Invite students to **share experiences pulling or riding in a wagon, or learning to ride a tricycle or bicycle.** What gives these vehicles their power? We do!

4. Talk with students about how **the wagon in this poem shows us everyday uses of energy, force, and motion**—using our hands and legs to pull the wagon (force) and then sending the wagon to roll away on its own (energy and motion). Students can push a crayon on their desks or a table and watch it roll over the edge (and then pick it up!).

5. Link this poem with another about moving a wagon, **"After I Made a Huge Mess with My Chemistry Set" by Mary Lee Hahn** (3rd Grade, Week 11, page 159), or follow up with the fun, classic story, *Go, Dog. Go!* by P. D. Eastman.

## Week 12: More Force, Motion & Energy

### Thank You, Isaac Newton
by **Eileen Spinelli**

My bookshelf falls upon the bed.
*Harry Potter* bonks my head.
Spaghetti slips—splat!—to the floor.
Clean-up is a messy chore.
Orange juice spills. Socks slide down.
Hail stones ping all over town.
Acorns plunk—ouch!—from a tree.
Oh, the joys of gravity!

### Take 5!

1. **Drop a book on the table or floor (with a firm *bonk*) as a clue to what the poem is about.** Then read the poem aloud, pausing briefly at the end of each line for added emphasis.

2. Ready for some noise? **Allow students to choose one small unbreakable item each (like a book) to drop on their desks or the floor as you read the poem aloud again.** They can drop that item when you say one of the sound words: *bonk, splat, plunk*. Then be sure they pick up each dropped item and talk about how *clean-up is a messy chore*.

3. **Do a bit of quick collaborative research on Isaac Newton.** Discuss why the poet included him in the title of the poem.

4. **Use this poem to talk about various objects and how they fall** (bookshelf, *Harry Potter* book, spaghetti, orange juice, socks, hail stones, and acorns). Do they fall down or up or back and forth? Fast or slow?

5. Share a poem that describes how the earth holds us close with **"Gravity" by Joyce Sidman** (2nd Grade, Week 11, page 119).

## Week 13: Light & Sound

### KINDERGARTEN

### Take 5!

1. **Vary the volume of your voice significantly as you read this poem aloud.** Read some lines in a normal tone, some in a loud voice (e.g., *but I'm everywhere*), and some in a whisper voice (e.g., *I vibrate soft whispers*).

2. Read the poem a second time and **invite students to echo each line after you read it, mimicking your volume for each line**—some loud, some soft, some normal.

3. Stop everything and encourage students to rest their heads, close their eyes, and **tune in to the sounds all around them for one minute.** Focus on the room, on the hall, outside—how far can they hear? Then identify the sounds and talk about how each sound travels on waves to our ears.

4. Help students understand the cause and effect relationship between sounds and hearing. **Talk about how this poem reminds us we can't SEE sounds,** but sounds, whispers, and music come to us on waves through the air. Students can cup their hands behind their ears to see how more sound waves can be captured.

5. Link this poem with another about noises all around us in **"Sound Waves at Breakfast" by Susan Marie Swanson** (2nd Grade, Week 13, page 121), and look for the Heinemann reader *What Is Sound?* by Charlotte Guillain.

## LISTEN
### by **Amy Ludwig VanDerwater**

I am a sound wave.
I travel through air.
No one can see me
but I'm everywhere.

I vibrate soft whispers
and songs in your ear.
I am a sound wave.
It's me that you hear.

## WEEK 14: SPACE

### KINDERGARTEN

### DID YOU KNOW?
by **Julie Larios**

. . . it snows metal on Venus?
That's what some scientists say.
One thing I wouldn't want
is to be there on a snowy day.

But one thing I *do* want—
one thing I've wished—
is to discover things like that
when I'm a scientist!

### Take 5!

1. Before sharing this poem, **invite students to tell what they know about the planets—including the featured planet, Venus.** Then read the poem aloud, pausing after the first line and before the final line for added impact.

2. The phrase *one thing* appears three times in the poem. **Invite students to say that phrase (*one thing*) each time it occurs** while you read the rest of the poem aloud. Cue them by raising your index finger each time.

3. For discussion: **What might you want to explore and discover as future scientists?**

4. **What fact do we learn about Venus in this poem?** Talk about how Venus is one of the planets that orbits the same sun that our planet Earth orbits. Show Venus on a map of the solar system.

5. For another space-themed poem, look for **"Uh Oh, Pluto!" by Jeannine Atkins** (2nd Grade, Week 14, page 122) or selections from Amy Sklansky's poetry book *Out of This World: Poems and Facts about Space*.

## Week 15: Sun, Earth & Moon

**KINDERGARTEN**

### Take 5!

1. If possible, **bring several balls of differing sizes as poetry props** to show while reading this poem aloud. Talk about how the sun and the stars are balls (or spheres) of varying sizes in space.

2. Read the poem aloud again, **pausing at the end of each line so that students can repeat each line, one at a time after you like an echo.**

3. **Discuss sun safety guidelines** such as not looking at the sun directly and wearing sunscreen.

4. **Do some quick collaborative research on the sun and its properties.** If possible, invite students to do a bit of sky study, identifying and describing what they see (sun, clouds, etc.).

5. Connect this poem with **"What I Know about the Sun" by Eileen Spinelli** (1st Grade, Week 15, page 83), and look for more space poems by Douglas Florian in his book *Comets, Stars, the Moon, and Mars.*

## Big Sun
by **Douglas Florian**

The sun seems BIG,
The BIGGEST star,
But that's because
It's near, not　　far.
And farther stars,
Though large in size,
Seem just a whisper
To our eyes.

## Week 16: The Water Cycle

### Old Water
by **April Halprin Wayland**

I am having a soak in the tub.
Mom is giving my neck a strong scrub.

Water sloshes against the sides.
$H_2O$'s seeping into my eyes.

The wet stuff running down my face?
She says it came from outer space!

The water washing between my toes
was born a billion years ago.

### Take 5!

1. As you read the poem aloud, **pantomime the actions suggested in the poem,** scrubbing your neck, wiping your eyes, wiping your face, lifting your feet or toes.

2. Read the poem aloud again and **invite students to join you in pantomiming the poem actions.**

3. Work with students to **identify all the words the poet uses to describe water:** *soak, water, slosh, $H_2O$, seep, wet, washing.*

4. **Discuss the water cycle briefly (how it *was born a billion years ago*) as well as examples of how water is useful to us in everyday life** (for baths, for cooking, etc.).

5. For another poem about the water cycle, look for **"Water Round" by Leslie Bulion** (2nd Grade, Week 16, page 124) and follow up with the nonfiction book, *Water Cycle* by Craig Hammersmith (part of the *Earth and Space Science* series).

# Week 17: Weather & Climate

**KINDERGARTEN**

## Take 5!

1. **If you have a dog or puppy puppet, use that to give a "face" to the poem as you read it aloud.** (Or make a simple dog puppet by holding up a magazine picture or online image of a dog.)

2. Read the poem aloud again, and this time **invite students to chime in on the thundering word *Kaboom*.**

3. For discussion: *How do pets and other animals act when there's stormy weather?*

4. Discuss the role of weather reporting in current events and how it helps us keep up with changes in weather from day to day. **Talk with students about the recognizable weather patterns that are described in this poem:** hot air, humidity, clouds, cooler weather, thunder, lightning, rain. Challenge them to describe today's weather and how it may have changed from yesterday.

5. Follow up with **X. J. Kennedy's poem** about weather and science, **"Discovery"** (1st Grade, Week 5, page 73), or selections from Linda Ashman's dog poetry book *Stella, Unleashed*.

## Dog in a Storm
### by Stephanie Calmenson

Yo! It's me—Buster the Brave.
Wait. I feel a storm coming.
The air is hot. It's humid.
Winds are blowing.
Clouds are rolling in.
The air is suddenly getting cooler.
KABOOM!
Thunder! Lightning! Rain, rain, rain!
I'm scared. I dive under the bed.
The weather reporter says,
"Thunderstorms may come when
cold air pushes warm air up—"
KABOOM! KABOOM! KABOOM!
Then the sky gets lighter.
The world gets quieter.
The rain stops.
I come out from under the bed.
Yo! It's me again—Buster the Brave.

# KINDERGARTEN

## WEEK 18: FORCES OF NATURE

### RIDDLE FOR A DRY DAY
by **Irene Latham**

Sun without rain
day in, day out.
Grass browns, ground frowns.
I am a DROUGHT.

### Take 5!

1. **Before sharing this poem, explain what a riddle is** (a guessing game with clues and an answer). Then read the poem aloud, reading the final line especially slowly, enunciating each word.

2. Talk about what a "drought" is (extremely dry land conditions) and practice the final line (*I am a DROUGHT*) with students. **Then read the poem aloud again, pausing before the final line for the children to chime in on the final line** in unison.

3. For discussion: *What are the good things and bad things about rain?*

4. **What weather details does the poet include in this poem?** (Examples include: sun, no rain for many days, brown grass, "frowning" ground, drought.) Talk with students about each of these descriptive words/phrases, showing images or pictures, if possible. Track weather changes for a week for comparison.

5. Contrast this poem with another by Irene Latham, **"Riddle for a Wet Day"** (1st Grade, Week 18, page 86), and look for more poems about the sun and rain in *Weather: Poems for All Seasons* compiled by Lee Bennett Hopkins.

## Week 19: Soil & Land

**Kindergarten**

### Take 5!

1. **The perfect prop for this poem? A rock, of course.** Read the poem aloud with the rock in your hand. If you don't have a rock that fits the poem exactly, talk about the differences between your rock and the rock in the poem.

2. For another reading of this poem, **coach students in memorizing the final line, *Are rocks living?*,** and invite them to chime in after you read the rest of the poem aloud.

3. Using your rock prop (or a picture of a rock), **invite students to come up with more descriptive words** about its size, color, shape, texture, weight, and so on.

4. Using the words in this poem (*cold, gray, white, hard, small, rough, round*), **talk about how scientists observe, describe, sort, and even compare objects like rocks.** How do they decide if an object is living or nonliving (identifying basic needs and the production of offspring)? Ask them to name some living and nonliving things in the room.

5. Revisit **"I Have a Question" by Anastasia Suen** (Kindergarten, Week 3, page 31), which also poses questions and wonders about answers. And look for the nonfiction companion book *Rocks: Hard, Soft, Smooth, and Rough* by Natalie Rosinsky.

## My Rock
### by Ken Slesarik

My rock is cold,
gray, white, hard,
small, rough
and round.
Are rocks living?

# KINDERGARTEN

## WEEK 20: NATURAL RESOURCES

### AUNTIE V'S HYBRID CAR
by **Janet Wong**

Auntie V drives a hybrid car.
It has a little screen
that shows green leaves
sprouting on a tree
when she's driving electric.
If she goes too fast,
she uses gas,
and leaves fall off.
Go slow, Auntie V—
and grow a tree for me!

### Take 5!

1. Before reading this poem aloud, **survey students on how many of them have seen or been driven in a hybrid car.** If necessary, explain that they use electricity as well as gasoline for their power.

2. Challenge students to participate in a follow-up reading of the poem by **saying the line *Go slow, Auntie V*** while you read the rest of the poem aloud. Practice the line first for added confidence.

3. **Discuss different modes of transportation** students and their families use (like walking, biking, cars, buses, etc.) and what energy sources each relies on.

4. **Talk with students about how different cars use different fuel sources.** What details does the poem provide about how hybrid cars work (informational screen, electric energy source, gas source)? Brainstorm other possible new or improved fuel-efficient cars or other modes of transportation that don't require fuel.

5. Connect this poem with another about how we power our cars with **"Fossil Fuels" by Janet Wong** (2nd Grade, Week 20, page 128). Also look for the nonfiction resource book, *What's So Bad about Gasoline?: Fossil Fuels and What They Do* by Anne Rockwell.

# Week 21: Ecosystems

## Kindergarten

### Alligator with Fish
by **Jane Yolen**

So many fish,
so many teeth,
dinner is always
just within reach.
Like a great colander,
like a big sieve,
the gator takes all
that the river will give.
He opens his mouth
in a big toothy smile.
Fish disappear.
He'll be full—
   for a while.

## Take 5!

1. **Show a picture of an alligator to set the stage** for reading this poem aloud. Search this helpful resource for various images of alligators: Animals.NationalGeographic.com.

2. **Time for movement!** Children can use their cupped hands to suggest fish swimming, then arms extended and clapping together for the jaws of the alligator as you read the poem aloud again.

3. For discussion: *What are your favorite foods to eat?*

4. This poem reminds us that animals prey upon one another for food. **Talk with students about the facts in this poem** about alligators and their need for a diet of fish, turtles, snakes, and/or small mammals for survival.

5. For another poem about the food cycle, look for **"A Biological Community/ *Una comunidad biológica*" by Margarita Engle** (4th Grade, Week 4, page 192). Or for more poems about various water creatures, seek out Douglas Florian's *In the Swim* and *The National Geographic Book of Animal Poetry* edited by J. Patrick Lewis.

# KINDERGARTEN

**WEEK 22: ADAPTATIONS & TRAITS**

## SNAKE TRAITS
by **Linda Ashman**

The slender,
hissing cobra
is quite famous for its bite.

But the wily anaconda
would much rather
squeeze you tight.

**Note:** The king cobra, found in southern Asia, is the longest venomous snake, growing up to 18 feet. If threatened, it spreads its hooded head and raises the front of its thin body to stand five feet tall. Although impressive, the cobra seems rather small compared to the anaconda of South America, which is the world's heaviest snake. This constrictor grows more than 30 feet long and can weigh close to 550 pounds.

### Take 5!

1. **Set the stage by sharing photos of snakes** from Seymour Simon's photo-essay book, *Snakes* (or from online resources). Then read the poem aloud with a long pause between stanzas.

2. Read the poem a second time **and invite students to hissssss with you** when you say the word *hissing* in the first stanza. **Then invite them to squeeze themselves tight** when you end the second stanza.

3. **Talk about the names for snakes** that students know, starting with *cobra* and *anaconda*.

4. Make a simple chart on the board and list the qualities provided in this poem for the cobra on one side (slender, hissing, biting) and for the anaconda on the other (wily, squeezing). **Talk about how scientists sort animals starting with physical characteristics, too.**

5. Look for **"The Lion and the House Cat"** by Mary Lee Hahn (1st Grade, Week 22, page 90) for another poem that compares two animals with distinctive similarities and differences.

*THE POETRY FRIDAY ANTHOLOGY FOR SCIENCE*

# KINDERGARTEN

## WEEK 23: CYCLES

### Take 5!

1. Do students know what radishes look like? If not, **show a picture of a radish or an actual radish** (if you can plan ahead). Then read the poem aloud slowly.

2. Since this poem is also presented in Spanish, **invite a Spanish speaker on your campus or in your community to read the Spanish version aloud.** Then read the English version aloud again. Record both readings for future reference, if possible.

3. For discussion: *What are your favorite vegetables to grow or eat?*

4. **Use this poem to identify the parts of a plant** including the seeds, leaves, stems, and roots. Note how the *tasty* parts that we usually eat are the *red/ roots* at the bottom of the radish. Consider other edible roots such as beets, turnips, or carrots.

5. For another poem about planting, revisit **"My Bean Plant" by Amy Ludwig VanDerwater** (Kindergarten, Week 7, page 35) or look for Lois Ehlert's picture book, *Planting a Rainbow* (also available in Spanish).

## YOUNG AND OLD TOGETHER
### by **Margarita Engle**

I love to help Grandpa in his garden,
planting tiny radish seeds
so we can watch the swift growth
of leaves and stems,
like green towers
on top of
tasty
red
roots.

## JÓVENES Y VIEJOS JUNTOS
### por **Margarita Engle**

Me encanta ayudar a mi abuelo en su jardín,
sembrando semillitas de rábano
para mirar cómo crecen tan rápido
las hojas y los tallos,
como torres verdes
encima de
sabrosas
raíces
rojas.

## WEEK 24: PATTERNS

### INHERIT TENSE
by **Charles Ghigna**

My family tree is rooted
Deep within my skin;
My eyes look like my mama's,
I have my daddy's chin.

I draw just like my grandma,
I sing like Uncle Lee.
The only question now remains—
What is left of me?

### Take 5!

1. **Display a family photo that includes multiple generations** (of your own or any family). Then read the poem aloud, pausing at the end of each line for added emphasis.

2. **Invite students to point to the body parts referenced in the poem** (*skin*, *eyes*, *chin*, and *me*) as you read the poem aloud again.

3. For discussion: *How are we like our parents or other family members? How are we different?*

4. Talk with students about how organisms resemble their parent organisms. **Use the poem to talk about physical features that children might share with their parents** (like skin color, eye color, facial features, tongue curling, etc.).

5. Look for another poem about family resemblances, **"Hand-Me-Downs" by George Ella Lyon** (1st Grade, Week 4, page 72).

## WEEK 25: HUMAN BODY

### KINDERGARTEN

### *Take 5!*

1. As you read this poem aloud, **use your hands to suggest the gestures mentioned in the poem** (bending, pushing, planting, catching, opening, pouring, holding, playing, throwing, and writing).

2. Read the poem aloud again, inviting students to do the gesturing and to **use their hands to pantomime the actions in the poem.**

3. For discussion: *If your hands are useful, what can you do with your feet?*

4. **Survey students on which of the poem activities are their favorites.** Make a quick graph of the results using pictures, numbers, and words.

5. Connect this poem with another poem about making things by hand, **"Driftwood Hut" by Renée M. LaTulippe** (1st Grade, Week 29, page 97). And look for the informational picture book *Hands Can* by Cheryl Willis Hudson.

## HANDS
### by **Kate Coombs**

I bend my fingers,
move thumbs like this.
Knuckles, nails,
palms and wrists.

With my two hands
I can push-pull hard.
I can plant a flower
in our backyard.

I can catch a bug,
I can open a door.
I can pour the milk
and touch the floor.

I can hold the baby,
play a video game.
I can throw a ball
and write my name.

My devices are nice—
they can talk, beep, sing.
But my two hands
can do anything.

## Week 26: Kitchen Science

### Can Our Eyes Fool Our Taste Buds?
by **April Halprin Wayland**

He loves the green drink.
She prefers red.

Guess what?

They're both the same!
The taste is in their heads!

**Note:** This poem was inspired by a psychology experiment that uses apple juice and red and green food coloring to examine whether people's perceptions of taste are influenced by their sight.
Check out: Education.com/science-fair/article/can-eyes-fool-taste-buds/

### Take 5!

1. **Hold up a glass, cup, or mug with water or a beverage in it as your poetry prop** as you read the poem aloud. Pause dramatically after the line *Guess what?*

2. When you read this poem aloud again, **coach students to read the one-line second stanza (*Guess what?*)** while you read the rest.

3. For discussion: *Have you ever been surprised by something not tasting the way you expected it to?*

4. If possible, **try conducting the experiment outlined in this poem.** Fill two glasses with apple juice and color one with red food coloring and the other with green food coloring. Invite another adult (teacher, parent, or staff) to try the two drinks and choose a favorite (if they can).

5. For another poem about investigations involving food coloring, revisit **"Capillary Action" by Joy Acey** (Kindergarten, Week 6, page 34).

## Week 27: Video Technology

**KINDERGARTEN**

### Take 5!

1. **If you have a cell phone, tablet, or computer with phone functions, use it as a prop** to set the stage for this poem before reading it aloud.

2. Invite the students to add to the technological "magic" of the poem by **saying the word *poof* in the poem as you read the rest of the poem aloud.**

3. For discussion: *What are the advantages and possible disadvantages of seeing people and being seen by them while talking on the phone?*

4. **If possible, arrange a Skype or FaceTime call** with another class, a colleague, or a family member (or a nearby principal or staff member). Show students **how we can use digital tools to communicate locally and globally to interact and collaborate with others.**

5. Link this poem with another about the telephone, **"Questions, Questions" by Ann Whitford Paul** (4th Grade, Week 3, page 191).

### HELLO, HELLO!
by **Janet Wong**

There's a button
on Mom's phone.
She presses it and—
poof!—
we're home,
sitting in the living room
with Grandpa asking,
"Where are you?"
Well, we could be
in Mexico
but you would never
ever know.
It feels like
we're down the hall
when we make
a video call!

## WEEK 28: MACHINES

**KINDERGARTEN**

### METAL MONSTER
by **X.J. Kennedy**

Mother lets out a loud cheer when
Our vacuum-cleaning robot
Purrs swiftly through our living room,
A metal monster. Oh, but

She sits and munches chocolates,
Watching it clean our rugs.
And when that sweet machine is done,
She gives it heartfelt hugs.

### Take 5!

1. **Set the stage for this poem by playing a short video** of a robotic vacuum cleaner such as this one of a cat and a Roomba: YouTube.com/watch?v=cLMj19FisJg. Then read this poem aloud.

2. Read the poem aloud again **and invite students to make sound effects in the background** that fit the lines of the poem, such as the mother's cheer, the robot purring, the mother munching, and the robot cleaning. Record the reading to enjoy again.

3. For discussion: *What chore around the house would you like to have a robot do for you?*

4. Talk with students about how the vacuum cleaner in this poem is special: it purrs, it is swift, it cleans, and it gets hugged. **Help them understand that a self-propelled vacuum cleaner is a type of robot.**

5. Partner this poem with another about a robot, **"My Robot" by David L. Harrison** (2nd Grade, Week 29, page 137). Compare both of these with the creations in *Robots, Robots, Everywhere!* by Sue Fliess.

# KINDERGARTEN

## WEEK 29: BUILDING THINGS

### Take 5!

1. **If possible, gather as many of the items mentioned in the poem as possible beforehand** *(gears, wheels, wire, twine, nuts, bolts, hooks, build-it books)*. Then place them in front of you and show each one as you read the poem aloud.

2. Read the poem aloud again and **invite students to chime in when the title of the poem appears within the poem** (*Tinker Time*). Cue them by holding up one of your poetry prop items.

3. Talk with students about **things they have created** "from scratch."

4. **Use this poem to talk about the steps we might follow in making something new:** gathering materials, consulting books, looking things up online. What kinds of things could students imagine creating using the items listed in the poem? Refer to Wikihow.com, a multilingual how-to site for ideas.

5. Revisit another poem about working with a grandparent, **"Young and Old Together / Jovenes y viejos juntos" by Margarita Engle** (Kindergarten, Week 23, page 51), or one about working with your hands, **"Hands" by Kate Coombs** (Kindergarten, Week 25, page 53). Also look for the resource book *Martha Stewart's Favorite Crafts for Kids: 175 Projects for Kids of All Ages to Create, Build, Design, Explore*.

### TINKER TIME
by **Janet Wong**

In Grandpa's basement you can find
gears and wheels and wire and twine,

lots of nuts and bolts and hooks
and one whole shelf of build-it books.

If we need help during Tinker Time,
we go to the computer and look online.

What will we build when we're all done?
We don't know yet—that's half the fun!

# KINDERGARTEN

## WEEK 30: SCIENCE FAIR

### SCIENCE FAIR DAY
by **Eric Ode**

Look at that!
Whoosh!
Splat!
Gurgle.
Pop.
Hooray!
It's Science Fair Day!

### Take 5!

1. **Hold up a beaker or test tube before reading this poem aloud,** pausing at the end of each line for added impact.

2. **Invite students to chime in on the exclamation** *Hooray!* while you read the rest of the poem aloud. Cue them by pumping your fist in the air.

3. **Talk with students about the sound words in the poem** (*whoosh, splat, gurgle, pop*) and about how they suggest what can happen when conducting science investigations using batteries or balls, for example—all with adult supervision.

4. Use this poem to talk about creating big projects, like science fair projects, and what that involves. Consult resources like the book *Yikes! Wow! Yuck! Fun Experiments for Your First Science Fair* by Elizabeth Snoke Harris.

5. Share another poem with science sound effects, **"Meet Mr. Wizard" by George Ella Lyon** (3rd Grade, Week 6, page 154).

*The Poetry Friday Anthology for Science*

# Week 31: Famous Scientists

KINDERGARTEN

## Rachel Carson
### by Julie Larios

Loving the earth—
its deep sea water,
its wide blue skies—
she listened for the sound of birds,
hoping she could help them
keep singing, hoping
we would never have
a silent spring.

### Take 5!

1. Using an online resource like AllAboutBirds.org, **play a moment of birdsong or show a "bird cam" (live camera feed) to set the stage** for this poem. Then read the poem aloud quietly and slowly.

2. Read the poem aloud again, and this time **encourage students to chirp bird noises when you say the word *birds*.** Pause for added impact there.

3. **Give students a little background on who Rachel Carson** was and why she is important to know about. She was a scientist who studied nature closely and wrote an important book called *Silent Spring* that helped advance the global environmental movement.

4. **Using this poem as a prompt, talk about what we observe in our environment:** earth (land), water, sky, sounds of birds. If possible, take a moment to study these same elements in your own surroundings.

5. For another poem about this pioneering scientist, look for **"Dear Rachel Carson" by Mary Lee Hahn** (1st Grade, Week 21, page 89). Also share the picture book biography *Rachel Carson and Her Book that Changed the World* by Laurie Lawlor.

## Week 32: More Famous Scientists

### Kindergarten

### OCEAN EXPLORER SYLVIA EARLE
by **Leslie Bulion**

She walks deep down on the ocean floor,
Where no one has ever walked before,
Then jumps in a submarine to explore more.

She teaches us each to do our part,
We *will* protect our oceans. We're smart!
We can save planet Earth's true blue heart.

---

**Take 5!**

1. **Use pantomime to add drama to the oral reading of this poem.** Pretend you are slowly walking on the ocean floor as you read the first stanza. Then use gestures for the second stanza: thumb to chest (*our part*), tap forehead (*We're smart!*), hand on your heart (*true blue heart*).

2. Share the poem again by **inviting students to chime in on the phrase *We're smart!*** while you read the rest of the poem aloud.

3. For discussion: *What are some things we can do to help "save planet Earth?"* (Recycle, use less water, etc.)

4. Use this poem to talk about how **scientists study different things in the natural world and use tools like submarines to help in their investigations.** Share details and images from the work of Sylvia Earle found at Mission-blue.org.

5. For a contrasting poem about the ocean in the future, look for **"Moving to Atlantis City, 2112" by Steven Withrow** (3rd Grade, Week 36, page 184), or look for Seymour Simon's nonfiction picture book *Oceans*.

# WEEK 33: COMPUTERS

## KINDERGARTEN

### Take 5!

1. **Display a scene from a simple video game or app** and then read this poem, pausing at the end of each line.

2. Read the poem aloud again and **invite students to echo you after you read each line.** If you have someone who can read the lines in Spanish, try echo reading those lines, too.

3. For discussion: *What are some of your favorite video games and apps?*

4. **Talk with students about what a game designer does:** working with computers, programs, and animation to create new games. Explain terms like *animated, move, designer,* if necessary.

5. Pair this poem with another about someone who creates video games, **"Game Programmer" by Janet Wong** (2nd Grade, Week 27, page 135). Also look for the "Cool Careers" book *Video Game Designer* by Kevin Cunningham.

### COMPUTER GEEK
by **Carmen T. Bernier-Grand**

Draw an animated creature
    Make it move in a game
        You'll be a Game Designer
I'll play your games.


### COMPU-NERDO
by **Carmen T. Bernier-Grand**

Dibuja una caricatura animada
    Hazla mover en un juego
        Serás diseñador de juegos
Jugaré tus juegos.

## Kindergarten
### Week 34: Science Careers

**Dr. Lee**
by **Janet Wong**

Last week
I couldn't
see the board
in any
of my classes.

Dr. Lee
saw
right away
that I just need
some glasses!

When I grow up
I want to be
a doctor
just like
Dr. Lee!

### Take 5!

1. **If you have glasses or wear glasses, use them as your poetry prop to set the stage** for reading this poem aloud.

2. Share the poem again, and this time **invite students to chime in on the pivotal word *glasses*** at the end of the second stanza. Pause and cue students by holding up your glasses prop.

3. **Collaborate with students to make your own custom eye chart** using EyeChartMaker.com.

4. **Use this poem to talk about the data** that revealed the need for glasses (couldn't see the board in class). Scientists pay attention when something is "different," and then ask why and try to find the reason. We can do that, too, even in our own lives.

5. Pair this poem with another about discovering the need for glasses, **"Seeing School" by Kate Coombs** (1st Grade, Week 25, page 93). Or look for the fun pop-up picture storybook *Arlo Needs Glasses* by Barney Saltzberg.

## Week 35: Future Challenges

**Kindergarten**

### Take 5!

1. **Place a clear glass of water in front of you as your poetry prop.** Take a sip and then read this poem aloud.

2. This poem is all about WATER, from the title to the final line. **Invite students to wiggle their fingers and move their arms from side to side to suggest flowing water each time the word *water* appears in the poem** while you read the poem aloud slowly.

3. For discussion: *How can we tell when water is safe to drink and when it is not?* For more information, go to: thewaterproject.org/resources/water_pollution_filtration_experiments.asp.

4. **Work with students to make a list of ways that water is important and useful to us** (to drink, for baths, for swimming, for cooking, for watering plants, etc.), beginning with examples from the poem (*fill the cup*).

5. Match this poem with another about how water comes to us, **"Water Engineered" by Sara Holbrook** (2nd Grade, Week 34, page 142). And for another poetic look at water, seek out *All the Water in the World* by George Ella Lyon.

## WATER
### by Kate Coombs

Turn on the tap
and the water flows.
Does anyone know
where the water goes?

Turn on the tap
and the water comes.
Does anyone know
where the water's from?

*Water is clean
and water is cool,
living in rivers
and raining in pools.*

*Yet water can trickle
and wells can dry up
till there's nothing left
to fill the cup.*

Today there is water
when we turn the tap on.
But what will we do
when the water is gone?

*The Poetry Friday Anthology for Science*

## Week 36: Future Dreams

### Future Dreams Idea #63
by **Janet Wong**

What if
you could make
a pillow-backpack

and when you touched a button,

it would pop up into
a Giant Air Shell
to keep you safe in an earthquake?

Mom asks what I'm doing
with my old broken backpack,
a roll of foil,
duct tape,
all our plastic bowls,
a whole bag of marshmallows,
my *sticky fingers*,
and her tablet.

I say, *Research!*

### Take 5!

1. **Place a backpack in front of you as your poetry prop.** Then read this poem aloud, pausing dramatically before the final word, *Research!*

2. Read the poem aloud again, but this time **invite students to say the last word (*Research!*).**

3. For discussion: ***What does it mean to be an inventor?***

4. **Use this poem to point out how a variety of everyday items** (old broken backpack, roll of foil, duct tape, plastic bowls, bag of marshmallows, tablet) **are used to try to make something brand new** (a *Giant Air Shell*). Talk about how we can use ideas and imagination to create new, original products using a variety of resources.

5. Revisit **"Tinker Time" by Janet Wong** (Kindergarten, Week 29, page 57), another poem about creating something new out of something old. Or for contrast, share the folktale *Joseph Had a Little Overcoat* by Simms Taback.

# Poems for First Grade

# NGSS Science and Engineering Practices: First Grade

*These practices form the foundation of disciplinary literacy in science and integrate reading, writing, listening, and speaking skills from the language arts. Here we indicate which weekly poems emphasize which science and engineering practices at each grade level.*

| PRACTICE | POEM |
| --- | --- |
| Asking questions and defining problems | Weeks 2, 3, 14, 35 |
| Developing and using models | Weeks 9, 28, 33, 36 |
| Planning and carrying out investigations | Weeks 5, 6, 7, 26, 29 |
| Analyzing and interpreting data | Weeks 4, 10, 16, 32 |
| Using mathematics and computational thinking | Weeks 8, 34 |
| Constructing explanations and designing solutions | Weeks 13, 20, 23, 25 |
| Engaging in argument from evidence | Weeks 11, 21, 22 |
| Obtaining, evaluating, and communicating information | Weeks 1, 12, 15, 17, 18, 19, 24, 27, 30, 31 |

# First Grade

| | | |
|---|---|---|
| week 1 | Scientific Practices | How to Be a Scientist *by Amy Ludwig VanDerwater* |
| week 2 | Lab Safety | The Science Lab Pledge *by Deborah Ruddell* |
| week 3 | Ask and Ask Again | Backwards *by Janet Wong* |
| week 4 | Observations | Hand-Me-Downs *by George Ella Lyon* |
| week 5 | Predictions & Hypotheses | Discovery *by X.J. Kennedy* |
| week 6 | Investigations | Pumpkin Experiment *by Mary Lee Hahn* |
| week 7 | Data | Testing My Magnet *by Julie Larios* |
| week 8 | Tools of Science | Celsius Thermometer *by Renée M. LaTulippe* |
| week 9 | Matter | Our Truck *by Janet Wong* |
| week 10 | More Matter | Sugar Water *by Janet Wong* |
| week 11 | Force, Motion & Energy | Love Note to a Magnet *by Patricia Hubbell* |
| week 12 | More FM&E | Frisbee *by Glenn Schroeder* |
| week 13 | Light & Sound | Prism *by Amy Ludwig VanDerwater* |
| week 14 | Space | Looking at the Sky Tonight *by Janet Wong* |
| week 15 | Sun, Earth & Moon | What I Know about the Sun *by Eileen Spinelli* |
| week 16 | The Water Cycle | Life Cycle *by Charles Ghigna* |
| week 17 | Weather & Climate | Clouds *by Kate Coombs* |
| week 18 | Forces of Nature | Riddle for a Wet Day *by Irene Latham* |
| week 19 | Soil & Land | Magic Show *by Juanita Havill* |
| week 20 | Natural Resources | Recycling *by Susan Blackaby* |
| week 21 | Ecosystems | Dear Rachel Carson *by Mary Lee Hahn* |
| week 22 | Adaptations & Traits | The Lion and the House Cat *by Mary Lee Hahn* |
| week 23 | Cycles | Photosynthesis *by Marilyn Singer* |
| week 24 | Patterns | I Like that Night Follows Day *by April Halprin Wayland* |
| week 25 | Human Body | Seeing School *by Kate Coombs* |
| week 26 | Kitchen Science | First Science Project *by Lesléa Newman* |
| week 27 | Video Technology | Pieces *by Renée M. LaTulippe* |
| week 28 | Machines | Levers *by Michael Salinger* |
| week 29 | Building Things | Driftwood Hut *by Renée M. LaTulippe* |
| week 30 | Science Fair | My Project for the Science Fair *by Kenn Nesbitt* |
| week 31 | Famous Scientists | Da Vinci Did It! *by Renée M. LaTulippe* |
| week 32 | More Famous Scientists | The "Black Leonardo" *by J. Patrick Lewis* |
| week 33 | Computers | The Engineer *by Stephanie Calmenson* |
| week 34 | Science Careers | Geologist *by Betsy Franco* |
| week 35 | Future Challenges | I Want to Know Why *by David L. Harrison* |
| week 36 | Future Dreams | Everyday Astronaut / Un astronauta común *by Carmen Tafolla* |

*"Science is a ladder . . . **poetry is a winged flight**."*

❧  Victor Hugo  ☙

## Week 1: Scientific Practices

**First Grade**

### How to Be a Scientist
by **Amy Ludwig VanDerwater**

Wonder.
Ask.
Hypothesize.
Experiment.
Open your eyes.
Watch.
Now write.
What do you see?
Share
everything
you learn
with me.

### Take 5!

1. **Say the title of the poem and ask students to share their guesses and responses about "how to be a scientist."** Then read the poem aloud slowly, enunciating each word.

2. **Invite students to echo read the poem,** repeating each line after you read it aloud, until you get near the end. Then pause and treat the last four lines as one segment to be repeated (*Share / everything / you learn / with me*).

3. **Work together to make a class poster with the steps outlined in the poem** (*wonder, ask, hypothesize, experiment, watch, write, share*) for future reference. Create a visual icon for each step (e.g., a cloud, a question mark, an X, a beaker, eyes, a pencil, and hands).

4. **Use the lines of this poem to help students think about what a scientist does** (beginning with *wonder, ask, hypothesize, experiment, watch, write, share*). Talk about how hypothesizing, guessing, and sharing are all important parts of scientific problem solving.

5. For another poem about how scientists like to ask questions, look for **"Inquiry" by Cynthia Cotten** (3rd Grade, Week 3, page 151). Also connect with Larry Verstraete's informational and poetic picture book *S Is for Scientists: A Discovery Alphabet*.

## WEEK 2: LAB SAFETY

**FIRST GRADE**

### THE SCIENCE LAB PLEDGE
by **Deborah Ruddell**

Be curious and careful.
Be organized and orderly.
Be goggled.
Be gloved.
Be safe.

### Take 5!

1. **Before reading this poem aloud, hold up a pair of safety goggles** and ask students to guess what this poem is about.

2. Next, read the poem aloud again and **invite students to chime in on the repeated word *Be* that begins each line.** Cue students by holding up a word card with *Be* written on it.

3. For discussion: ***When are times it's fun or helpful to wear goggles or gloves?***

4. **Here is the teachable moment for talking about the importance of safety in science.** Research your local safety standards and discuss with students the value of **safety goggles**, washing hands, and using materials appropriately.

5. For another poem about the gear one needs to be safe with science, look for **"Superhero Scientist" by Joan Bransfield Graham** (Kindergarten, Week 2, page 30). Also look for the helpful nonfiction book *Science Safety Rules* by Kelli Hicks.

## Week 3: Ask and Ask Again

**First Grade**

### Take 5!

1. **Create a simple poetry prop** by cutting out a circle of construction paper or poster board and gluing it to a popsicle stick. Write the word "why?" on one side and the words "why not?" on the other. Then use the prop as you read the poem aloud, showing the side that corresponds to the poem and pausing briefly between stanzas.

2. In sharing the poem again, **students can say the words *Why* and *why NOT?* as they occur in the poem** while you read the rest aloud. Cue students by cupping your hand around your ear.

3. **Try a half hour or hour of "backwards" time**—read the end of a book, line up backwards Z-A, etc.

4. **This poem can launch a discussion of how asking questions and seeking answers is fundamental to scientific investigation and reasoning.** Start a question wall with craft paper on a door or bulletin board and invite students to brainstorm and add their own "why/why not?" questions to ponder.

5. Connect this poem with another about asking questions, **"I Have a Question" by Anastasia Suen** (Kindergarten, Week 3, page 31), or look for the intriguing, question-filled picture book *Ask Me* by Antje Damm.

## Backwards
by **Janet Wong**

Everyone is asking WHY.
*Why is the sun hot?*

I know I should wonder why but
I'm thinking:
*Well, why NOT?*

## WEEK 4: OBSERVATIONS

**FIRST GRADE**

### HAND-ME-DOWNS
by **George Ella Lyon**

I look like my mother.
My brother looks like my dad.
Mom looks like her mother.
Dad looks like Grandpa Tad.

Mom says this is genetic
but I don't know what that means.
She says that I'll learn someday
when I study genes.

I look at the ones I'm wearing.
I look at my brother's too.
I can't wait till I am old enough
to learn what jeans can do.

### Take 5!

1. Write the words "genes" and "jeans" on the board before reading the poem aloud. **Then talk about the difference between the two words as indicated in the poem** (*genes* help create our bodies and personalities; *jeans* are pants that we wear).

2. Read the poem again, and this time **invite students to chime in on the words "genes" and "jeans"** while you read the rest of the poem aloud.

3. **Talk about the expression "hand-me-downs" and invite students to talk about things they have shared with or received from** family members or friends (e.g., favorite item of clothing, shoes, toys, books, games, etc.).

4. If possible, **share a photo of your own parents and challenge students to note how you look like them and not like them** to encourage their observation skills. Make a simple table with two columns showing both sides (like/not like) to communicate their findings.

5. For another poem about what we get from our family at birth, share **"Inherit Tense" by Charles Ghigna** (Kindergarten, Week 24, page 52), and seek out *Take Two! A Celebration of Twins* by J. Patrick Lewis and Jane Yolen.

## Week 5: Predictions & Hypotheses

### Discovery
by **X.J. Kennedy**

Benjamin Franklin from
Old Philadelphia
Sent up a door key tied fast to a kite,
Guessing that lightning was sheer electricity.
Turned out that wise guy
Was perfectly right.

### Take 5!

1. **Set the stage for this poem by showing a few pictures of Benjamin Franklin** (search Biography.com). Then read the poem aloud slowly and talk briefly about Franklin's experiment with kites, keys, and electricity.

2. For a second reading, coach students to say the line *Was perfectly right*. Read the poem aloud again, then pause to cue them for their final unison line.

3. Discuss lightning safety with students. Look for lightning safety myths and facts at LightningSafety.noaa.gov/facts_truth.htm.

4. **Talk about the tools this poet mentions in describing Benjamin Franklin's experiment** (*door key, kite*). Scientists investigate different things in the natural world and use all kinds of tools to help in their investigations.

5. Make a "kite" connection with **Laura Purdie Salas's poem "Go Fly a Kite"** (2nd Grade, Week 12, page 120). Or look for a simple, highly visual biography for young children such as *Time for Kids: Benjamin Franklin: A Man of Many Talents* or *How Ben Franklin Stole the Lightning* by Rosalyn Schanzer.

## WEEK 6: INVESTIGATIONS

### FIRST GRADE

### PUMPKIN EXPERIMENT
by **Mary Lee Hahn**

We put one pumpkin
in the land lab—
left it there
all fall long.

It shrank and shriveled
in the land lab—
after winter,
it was gone.

Where we left it
in the land lab—
pumpkin plants
are growing green.

Five fat pumpkins
in the land lab—
four to carve
and one for seeds.

**Note:** Our land lab is an outdoor laboratory at our school.

### Take 5!

1. **Set the stage for this poem with the wordless video** "Time Lapse Pumpkin Vines Growing" at YouTube.com/watch?v=DJGJiAKS90M. Then read the poem aloud, pausing briefly between each stanza.

2. The second line, *in the land lab*, is repeated in every stanza in this poem. **Invite students to chime in on that line each time** as you read the rest of the poem aloud. Talk about what a "land lab" might look like.

3. Invite students to **share experiences they have had with planting, gardening or farming.**

4. **With this poem as an example, talk about the energy cycle described here, beginning** with the pumpkin in the fall and its process of rotting or decomposing in the winter. Research during which season there might be green plants showing and when there might be a pumpkin again.

5. Another poem that features the natural process of decomposition is **"Mold" by Charles Waters** (2nd Grade, Week 26, page 134). And share more information about pumpkins with the nonfiction books *Pumpkins* by Ken Robbins and *The Pumpkin Book* by Gail Gibbons.

## Week 7: Data

### First Grade

### Testing My Magnet
by **Julie Larios**

Flowers? No. Dirt? No.
Socks? No. Shirt? No.
Hamster? No. Snake? No.
Plastic scoop and rake? No.
Glue? Paint? Paper? Clay?
Sneakers that I wore today?
No, no, no, no . . .

Pile of metal paper clips—
Yes! Hooray for paper clips!
Shiny whistle? Metal fan?
Dented metal garbage can?
Hammer head, bag of nails?
Ring of keys? Rusty pails?
Yes, yes, yes, and yes!

Results of my experiment?
Magnets are mag-nificent!

### Take 5!

1. **If you have a magnet handy, now is the time to use it as your poetry prop** to demonstrate some of the examples mentioned in the poem as you read aloud.

2. Share the poem again, and this time **invite students to chime in with the appropriate response, *no* or *yes*,** as these words repeatedly appear in the poem. Cue them with a thumbs up or thumbs down signal.

3. For discussion: *Is it better to test one thing or many things? Why?*

4. Show how this poem is a collection of observations and data about what is and is not magnetic. Count how many things are tested. The first stanza features items that are NOT magnetic, the second stanza features things that ARE magnetic. **Challenge students to predict what other objects might or might not be magnetic** and, if possible, test their predictions.

5. Link this poem with another about how magnets work, **"No Penguins Here" by Michael Salinger** (5th Grade, Week 11, page 239), or connect with another poem that presents data and results, **"Zapped!" by April Halprin Wayland** (3rd Grade, Week 7, page 155).

## WEEK 8: TOOLS OF SCIENCE

### CELSIUS THERMOMETER
by **Renée M. LaTulippe**

The teacher taught us Celsius
to measure temperature.
She showed us the thermometer
and where the numbers were.

I learned that water's freezing point
is right at 0 degrees.
100 makes the water boil—
Celsius is easy!

### Take 5!

1. Write the numbers 0 and 100 on the board and read the poem aloud. Then add to your notation: *0 = freezing* and *100 = boiling*. **Talk about how Celsius is a different measurement system than we usually use** (Fahrenheit).

2. In sharing the poem aloud again, **students can say the key word, *Celsius*, as it occurs in the poem.** Cue students by raising your hand in the shape of a C.

3. **Survey students about which they like better, hot weather or cold weather,** and make a list of their reasons in two columns.

4. This poem reminds us that a thermometer can be a helpful measurement tool for science. **Talk with students about how Celsius uses 0 (freezing) and 100 (boiling), in contrast with Fahrenheit,** which uses 32 (freezing) and 212 (boiling). If possible, show a thermometer with either or both scales. Consider how water changes at different temperatures and how we can measure that two different ways.

5. Connect this poem with another about the metric system, **"Meter Stick" by Amy Ludwig VanDerwater** (2nd Grade, Week 8, page 116), or with another poem about measurement, **"Rain Gauge" by Anastasia Suen** (2nd Grade, Week 17, page 125).

## Week 9: Matter

**First Grade**

### Take 5!

1. While you read this poem aloud, **in the background show a time lapse video of a car rusting** (without the audio). One source is Wonderopolis.org/wonder/why-do-some-things-rust/.

2. Share the poem again and **invite students to chime in on the key lines,** *water and air* and *the iron in the steel,* as you read the poem aloud.

3. **Work together to brainstorm a list of everyday objects made of iron or steel** (e.g., bike, tin can, fire extinguisher, lock, horseshoe). See: Sciencephoto.com/media/221347/view.

4. **Use this poem to talk about why things rust** *(water and air / mixing with / the iron in the steel).* Make a chart of things that can rust and things that cannot rust. Again, consult Wonderopolis.org/wonder/why-do-some-things-rust/.

5. For another poem about things made of steel, look for **"Computer Models" also by Janet Wong** (4th Grade, Week 8, page 196), or for another poem about trucks, seek out **"Sound Waves at Breakfast" by Susan Marie Swanson** (2nd Grade, Week 13, page 121).

## Our Truck
### by Janet Wong

Last year
our truck
got busted up—
and we couldn't fix it.
Now it's covered in rust.
Dad says it's just
water and air
mixing with
the iron in the steel.
And the paint
that used to be so smooth
is now as bumpy
as an orange peel!

## WEEK 10: MORE MATTER

**FIRST GRADE**

### SUGAR WATER
by **Janet Wong**

We dump some lumps of sugar
in cold water. Stir, stir, stir.

There's sugar at the bottom
though we stir, stir, stir, stir, stir.

What happens with hot water?
The sugar disappears!

Excuse me, Sugar:
are you still here?

### Take 5!

1. **If you have a cup, mug or pitcher, and a spoon handy, use this as your prop** while reading the poem aloud. Stir, stir, stir the empty cup for added sound effects.

2. In sharing the poem again, **students can say the word *stir* each time it occurs** while you read the rest of the poem aloud. Cue students by vigorously stirring your spoon in your mug (and not the rest of the time).

3. Check out the U.S. Geological Survey's Water Science School online at GA.water.usgs.gov/edu/, where **students can test their water knowledge** (in Spanish and Chinese, too!).

4. **Here is a mini-experiment in a poem. If you can demonstrate it briefly** with hot and cold water and sugar packets, go for it. Talk with students about what sugar looks and tastes like and how it changes when it is in water, depending on the temperature of the water, cold or hot.

5. For another poem mini-experiment with a beverage, look for **"Can Our Eyes Fool Our Taste Buds?" by April Halprin Wayland** (Kindergarten, Week 26, page 54). Or for another poem about mixing with water, look for **"Water + Dirt =" by Rebecca Kai Dotlich** (Kindergarten, Week 9, page 37).

## Week 11: Force, Motion & Energy

**First Grade**

### Love Note to a Magnet
by **Patricia Hubbell**

Dear Magnet,
I'm drawn to you.
Irresistible you!
You're so attractive.
You make me feel quite *active*!
And though I know I should shrug,
I can't resist the tug of this feeling of love.
I want to hop, jump, and skip,
Really let rip . . .
So I'm on my big trip—
Past pens, pencils, and papers,
Rulers, tape, and erasers—
Because . . .
I'm drawn to you!
Hugs,
        Paper Clip

### Take 5!

1. Add a bit of fun to sharing this poem with a poetry prop: **show a paper clip before reading the poem aloud.**

2. **Next, divide students into two groups**—one to say the word *Magnet* and one to say the closing phrase, *Paper Clip*. Read the poem aloud again, and if you have a magnet and a paper clip handy, use those to cue students to participate.

3. For discussion: *How are magnets helpful in everyday life?*

4. **If possible, show how a magnet can be used to pull an object.** Demonstrate how a paperclip might be magnetic (unless it's plastic), and then test other school supplies to see if they are magnetic.

5. Revisit a previous related poem, **"Testing My Magnet" by Julie Larios** (1st Grade, Week 7, page 75) for comparison, and look for the book *School Supplies* compiled by Lee Bennett Hopkins for more poems like these.

## WEEK 12: MORE FORCE, MOTION & ENERGY

**FIRST GRADE**

### FRISBEE
by **Glenn Schroeder**

Why does a Frisbee go far with a fling?
Because it moves air like an airplane wing.

Why does a Frisbee fly flat and not wiggle?
Because it is spinning, which stops the jiggle.

**Note:** Frisbee is a registered trademark of Wham-O.

### Take 5!

1. Before reading this poem aloud, **be sure students know what a Frisbee is** and show one if possible (FrisbeeDisc.com). Then read the poem aloud, pausing at the end of each line for emphasis.

2. In repeated readings, **coach students to say the opening phrase *Why does a Frisbee*** while you read the rest of the poem aloud. Cue them by cupping your hand behind your ear.

3. **Talk about other objects that can "fly."**

4. **Use this poem to talk about how a Frisbee moves through the air**—flat, without a *wiggle* or *jiggle*, because it is *spinning* while flying. If possible, throw a Frisbee outdoors to demonstrate. Indoors, a sturdy paper plate or pie tin flung upside down can be a good substitute.

5. For another poem about fun with flying objects, look for **"Go Fly a Kite" by Laura Purdie Salas** (2nd Grade, Week 12, page 120), and don't miss *Make Things Fly: Poems about the Wind* edited by Dorothy Kennedy.

## WEEK 13: LIGHT & SOUND

**FIRST GRADE**

### Take 5!

1. **Start with images from Optics4kids.org** (look in the "Gallery") to set the stage before reading this poem aloud.

2. Then read the poem aloud again, and **invite students to chime in on their favorite color** (*red, orange, yellow, green, blue, indigo, violet, white*) while you read the rest of the poem aloud.

3. **Survey students about their favorite colors.** Make a quick graph of the results.

4. Talk with students about how the **ROY G BIV colors** (red, orange, yellow, green, blue, indigo, violet) **hide in white light,** and about how different forms of energy such as light are important to everyday life. Brainstorm a list of ways that light is useful in the classroom, library, and home.

5. Pair this poem with another about how light is important, **"To the Eye" by Laura Purdie Salas** (4th Grade, Week 13, page 201). Also look for Joan Bransfield Graham's book of concrete poetry, *Flicker Flash*.

## PRISM
### by Amy Ludwig VanDerwater

White light holds a treasure.
A prism is the key.
I hold it up to sunshine—

Colors!
You are free!
Red and Orange.
Yellow, Green, Blue.
Indigo.
Violet.
Let me look at you!

Bend a rainbow
from white light.
You've been hiding
in plain sight.

## WEEK 14: SPACE

### LOOKING AT THE SKY TONIGHT
by **Janet Wong**

Dad and I look up
at the sky.
He points to dots
and asks what they are.
I say "well . . . stars,"
but he wants
something else,
some different words.
He's tracing a shape
with his finger in space—
and now I see!
It's a measuring cup
or a powder drink scoop—
or something bigger—
smaller than a pot—
but not by a lot—
how about we call it
The Big Dipper?

### Take 5!

1. **Poke a hole in a piece of black paper with a pin. Hold this up as your poem prop** for scale as you read this poem aloud slowly, pausing at the end of each line for emphasis.

2. **Next, divide the students into two groups**—one to say the phrase *well . . . stars* and one to say the phrase *The Big Dipper*. Cue them by raising either one or two fingers.

3. For discussion: *When you look up in the night sky, what do you imagine?*

4. **Talk with students about how stars and planets in space look like "dots" to us on earth** and how the natural world includes the air around us and objects in the sky. Use the resources at Space.com (Space Photos) for visuals, particularly the "Image of the Day."

5. For another poem about entities in space, share **"Orion Nebula" by Mary Lee Hahn** (3rd Grade, Week 14, page 162), and look for the *National Geographic Kids' First Big Book of Space* by Catherine D. Hughes for more interesting information.

## WEEK 15: SUN, EARTH & MOON

**FIRST GRADE**

### Take 5!

1. Before sharing this poem, **ask students to share three things they know about the sun.** Invite them to listen for more sun facts as you read the poem aloud.

2. Share the poem again and **invite students to chime in on the repeated phrase *I know*** while you read the rest of the poem aloud. Cue them with the ASL sign language gesture for "I know": tap your forehead twice with the tips of your fingers of your outstretched hand. Find video help at SigningSavvy.com.

3. For discussion: **How are the sun and the moon alike and different**? (Both are in space and give off light; one is visible mainly in the day and one mainly in the night, etc.)

4. **Use this poem to talk about the functions and attributes of the sun** (e.g., it is a star millions of miles away and helps things grow; it takes eight minutes for its light to reach the earth; it is always shining). If possible, take a moment to observe the sun in the sky and talk about those observations. Discuss sun safety guidelines such as not looking at the sun directly and wearing sunscreen.

5. Connect this poem with another about the sun, **"Big Sun" by Douglas Florian** (Kindergarten, Week 15, page 43), and look for the nonfiction reader *The Sun* by Melanie Chrismer for even more information.

### WHAT I KNOW ABOUT THE SUN
by **Eileen Spinelli**

I know that the sun is a dazzling star
far, far from earth. Millions of miles far.
I know that plants, animals, and people
need the sun to grow.
I know that it takes eight minutes or so
for the light of the sun to reach earth.
And that the sun is always shining somewhere
even when it's dark in my back yard.
I also know how the sun shimmers on the pond
where my grandpa takes me fishing.
And how quickly it bakes mud pies on an August day.
I know how the sun brightens everything—even hearts.
And how poets like to sing about it.

## Week 16: The Water Cycle

**First Grade**

### Life Cycle
by **Charles Ghigna**

The stream
becomes
the river
becomes
the root
becomes
the tree
becomes
the sky
becomes
the cloud
becomes
the rain
becomes
the stream

### Take 5!

1. Feeling creative? Take a marker, crayon, or piece of chalk (whatever medium is most handy) and **draw each object in the poem slowly with one continuous line** (like *Harold and the Purple Crayon*) as you read the poem aloud: *stream-river-root-tree-bird-sky-sun-cloud-rain-stream*. It doesn't have to be perfect; it just has to suggest the connectedness of all these elements.

2. For another experience, **invite students to say the repeated word *becomes*** while you read the rest of the poem aloud.

3. For discussion: ***What do you want to become when you grow up?***

4. **Use this poem to talk about how the natural world includes rocks, soil, and water** that can be observed in cycles and patterns. Talk about how each element is connected and what might happen if the connection is broken (through pollution, for example, or if the weather is too rainy or too dry).

5. For another look at the water cycle, seek out **"Water Round" by Leslie Bulion** (2nd Grade, Week 16, page 124) or **"Old Water" by April Halprin Wayland** (Kindergarten, Week 16, page 44). Also seek out Kate Coombs's book of ocean poems, *Water Sings Blue*.

## Week 17: Weather & Climate

**First Grade**

## Clouds
### by Kate Coombs

I saw one little cloud
that looked like a wish,
but now there's a crowd
like a school of white fish.

Clouds can turn red at sunset
or shine with gold light.
Sometimes dark clouds growl
with thunder at night.

There are clouds flat as paper
and clouds fat as geese,
clouds built like staircases,
clouds soft as fleece.

But clouds *should* look wet—
and do you know why?
All clouds are secretly
lakes in the sky.

### Take 5!

1. Tear a piece of white paper all around the edges or stretch and glue cotton balls to paper to **create a cloud prop.** Then read the poem aloud, pausing briefly between stanzas.

2. Read the poem aloud again, and **invite students to say the word *cloud* each time it pops up** in the poem. Cue them by holding up your homemade cloud.

3. **Take a moment to look outside, see what today's clouds look like, and share impressions.** Or for inspiration, share the classic picture book *It Looked Like Spilt Milk* by Charles Shaw.

4. **Use this poem to prompt a cloud study.** Challenge students to observe and record changes in the appearance of clouds in the sky for one week. *Clouds* by Anne Rockwell can be a helpful resource book for identifying different types of clouds and the weather associated with them. Or check out the CloudAppreciationSociety.org.

5. Connect this poem with **"Tropical Rain Forest Sky Ponds" by Margarita Engle** (4th Grade, Week 21, page 209) about a forest in the sky. Also look for Rita Gray's poetry book *One Big Rain: Poems for Rainy Days*.

## WEEK 18: FORCES OF NATURE

### RIDDLE FOR A WET DAY
by **Irene Latham**

I overwhelm and overflow
with raging waves and sheets of mud.
I leave behind disease and sludge.
I am an unexpected FLOOD.

### Take 5!

1. Before sharing this poem, ask students if they know what a riddle is (a word puzzle with clues and an answer). Invite them to listen for the clues in this poem and wait for them to guess the answer as you read the poem aloud.

2. Read the poem aloud again, and **invite students to say the last word together**—the answer to the riddle.

3. For more poem riddles, share selections from *Spot the Plot! A Riddle Book of Book Riddles* by J. Patrick Lewis.

4. **Use this poem as a prompt to talk about how rain and floods can affect the soil** (e.g., by unexpectedly creating mud and mudslides, disease, and sludge). Work together to research news stories about recent floods as examples.

5. Look for the contrasting poem by the same poet, **"Riddle for a Dry Day" by Irene Latham** (Kindergarten, Week 18, page 46). Also find *Weather: Poems for All Seasons* edited by Lee Bennett Hopkins.

## WEEK 19: SOIL & LAND
### FIRST GRADE

### MAGIC SHOW
by **Juanita Havill**

I don't use a wand.
I don't use a hat,
nor bright silk scarves.
All I need is a vat.

Into my vat—
with slits on the sides—
go corn husks, dead plants,
leaves that have dried,

tea bags, coffee grounds,
veggie peels galore,
old straw, sawdust,
egg shells, and more.

Water if needed.
Be sure to add dirt.
Turn with a pitchfork.
Better than dessert.

No bunnies or doves
in this magic show.
Abracadabra!
Let the compost flow.

### Take 5!

1. After reading this poem aloud, **you may need to explain a few key words** like *vat* and *compost* that are essential to understanding the poem.

2. Invite students to select their favorite compostable item and to **chime in on that word or phrase while you read the rest of the poem aloud.** These include: *corn husks, dead plants, leaves that have dried, tea bags, coffee grounds, veggie peels galore, old straw, sawdust, egg shells, water, dirt.*

3. For discussion: **What kinds of "garbage" can decompose in soil and which cannot?**

4. If possible, **bring in a sample of soil to observe and describe in terms of size, texture, and color.** Talk about the role of composting in creating fertilizer for soil. Look for educator resources at the CompostingCouncil.org.

5. For another poem on decomposition, revisit **"Pumpkin Experiment" by Mary Lee Hahn** (1st Grade, Week 6, page 74) and look for more poems by Juanita Havill in *I Heard It from Alice Zucchini: Poems About the Garden.*

## Week 20: Natural Resources

**First Grade**

### Recycling
by **Susan Blackaby**

Collect the daily scraps and clippings,
gather up the bits and snippings:
Paper, plastic, glass, and tin—
all of these go in the bin.
Once it's sorted and inspected,
so-called waste is redirected.
Think of all the things that you
can make from useful stuff you threw
away!

### Take 5!

1. **Pile some trash in front of you as your poetry prop,** including some items that could be recycled. Then read the poem aloud.

2. Read the poem aloud again, and **invite students to chime in on the words and phrases that describe what can be recycled:** *scraps and clippings; bits and snippings; paper, plastic, glass, tin.* You read the rest of the poem aloud.

3. For discussion: **What kinds of things do you recycle at home?**

4. **Guide students in identifying recyclable items** mentioned in the poem (*paper, plastic, glass, tin*) and beyond and find examples of these in your classroom, showing children the recycling symbol on items that contain it. Gather one day's worth of scrap paper in a pile to demonstrate how small amounts of paper can add up.

5. Revisit last week's poem about the environmentally responsible practice of composting, **"Magic Show" by Juanita Havill** (1st Grade, Week 19, page 87). Just for fun, check out *The Green Mother Goose: Saving the World One Rhyme at a Time* edited by Jan Peck and David Davis.

## WEEK 21: ECOSYSTEMS

### DEAR RACHEL CARSON
by **Mary Lee Hahn**

Dear Rachel Carson,

We went to the organic farm yesterday
and learned about you, and why they don't spray
chemicals to kill bugs that eat up their crops:
the balance in nature goes from bottom to top.

You warned in your book, *Silent Spring*, long ago,
humans must always be sure that they know
what the impact will be on *all* living things
when we do things to benefit us, human beings.

Someday when we're scientists, we'll think of you
and remember your teaching in all that we do.
In our work to help humans we'll never forget
that we're only one part of the life on our planet.

Yours truly,
Miss Smith's 1st Grade Class

### Take 5!

1. **Project an image of Rachel Carson while you read this poem aloud.** One resource with slide images is RachelCarson.org.

2. **Use the letter format of this poem to invite student participation.** They can read the opening (*Dear Rachel Carson*) and the closing (*Yours truly, / Miss Smith's 1st Grade Class*) while you read the rest of the poem aloud.

3. **Give students a little background on who Rachel Carson** was and why she is important to know about. She was a scientist who studied nature closely and wrote an important book called *Silent Spring* that helped advance the global environmental movement.

4. **Use this poem to talk about how Carson studied the environment,** *the balance in nature,* and *the impact on all living things.* Consider the interdependence among living organisms (e.g., *chemicals to kill bugs that eat up their crops,* but the chemicals can also affect the helpful bugs and other life there).

5. For another poem about this pioneering scientist, look for **"Rachel Carson" by Julie Larios** (Kindergarten, Week 31, page 59), as well as the picture book biography *Rachel Carson: Preserving a Sense of Wonder* by Joseph Bruchac.

## WEEK 22: ADAPTATIONS & TRAITS

**FIRST GRADE**

### THE LION AND THE HOUSE CAT
by **Mary Lee Hahn**

different strength
different size
same chin
same eyes

different mane
different stride
same stretch
same pride

### Take 5!

1. Before reading this poem aloud, **show images of a lion and a house cat** to see them side by side.

2. **Next, divide students into two groups**—one to say the word *different* as it occurs in the poem and one to say *same* each time it appears in the poem. Once again, cue them by raising your index finger high for *different* and extending your palm for *same*.

3. Talk about the details in the poem and **create a simple Venn diagram** showing the traits of the lion, those of the house cat, and those held in common.

4. **Work together to investigate** how the characteristics of the lion are related to where it lives, how it moves, and what it eats. Look for help at Kids.NationalGeographic.com.

5. For another poem of comparison (the cobra and the anaconda), look for **"Snake Traits" by Linda Ashman** (Kindergarten, Week 22, page 50). For more information about lions, look for *Face to Face with Lions* by Dereck Joubert, and for a fun poetry book about a house cat, look for *Won Ton: A Cat Tale Told in Haiku* by Lee Wardlaw.

## Week 23: Cycles

**First Grade**

## Photosynthesis
by **Marilyn Singer**

You start with lots and lots of air
(preferably clean),
the right amount of water,
and, oh yeah, you must be green.
But it won't work, you'll make no fuel,
try with all your might,
if you're not some kind of a plant,
and you haven't any light.

### Take 5!

1. To set the stage for this poem, **place a green plant in front of you** before reading the poem aloud. Afterward, talk briefly about the poem title and topic, photosynthesis, and how plants absorb sunlight and turn that energy into food.

2. In reading the poem aloud again, **invite students to read lines in unison, alternating with you as the leader.** You read the first line aloud, they read the second line, you read the third line, and so on.

3. **Talk about how plants are affected by the seasons of the year** (less light in winter, plants are dormant or die).

4. **Work together to sketch a diagram of the cycle of** photosynthesis detailed in this poem (clean air, water, green plant, light). Need help? Go to PhotosynthesisForKids.com.

5. Another poem that incorporates photosynthesis is **"Sun-Kissed / Besado por el sol" by Guadalupe Garcia McCall** (3rd Grade, Week 23, page 171), a poem filled with sun imagery. Add to that Alma Flor Ada's poetry book *Gathering the Sun: An Alphabet in Spanish and English*.

## WEEK 24: PATTERNS

### I Like that Night Follows Day
by **April Halprin Wayland**

What if there was never night—
if it was always light . . . and light?

No dark, no yawn, no closing eyes.
No moon or stars in any skies.

No quiet that the nighttime brings.
I'm sort of scared to think those things.

A sky that's dark as my dog's nose
is just the way things ought to go.

### Take 5!

1. Before sharing this poem, take a moment to **encourage students to close their eyes and think about being home at nighttime**. Then continue by reading this poem aloud.

2. Share the poem again, and **invite students to say the lines that highlight the qualities of the night**—the three lines that begin with the word *No*—while you read the rest of the poem aloud.

3. Use this poem **to identify and discuss the characteristics of day and night** (day = *light*; night = *dark, yawn, closing eyes, moon, stars, quiet*, etc.). For discussion: **What are things to like about each (day/night)?**

4. **Talk about the patterns that exist in nature:** day and night and the seasons of the year (spring, summer, fall, winter).

5. For a poem about the moon, look for **"Queen of Night" by Terry Webb Harshman** (4th Grade, Week 15, page 203), and seek out the Caldecott Award-winning picture book *The House in the Night* by Susan Marie Swanson.

## Week 25: Human Body

### FIRST GRADE

## SEEING SCHOOL
by **Kate Coombs**

My desk is in back
and I can't really see.
I ask to move up—
now it's glasses for me.

Don't want to wear them.
*Do I have to, Mom?*
But the world takes shape
when I put them on.

I see letters and words
all over the place,
numbers and edges
and my teacher's face.

I can see all the smiles
when I come to class.
Today I'm brand-new
with two pieces of glass.

### Take 5!

1. **What is the perfect prop for this poem? A pair of glasses, of course.** Put on a pair of glasses (propped on top of your head, if best for reading) and read this poem aloud.

2. **Since this poem has a line of dialogue, invite students to say the line** *Do I have to, Mom?* while you read the rest of the poem aloud.

3. **Collaborate with students to make your own custom eye chart** using EyeChartMaker.com.

4. **Use this poem to talk about the evidence** that revealed the need for glasses and how vision was improved after wearing glasses (can't see from the back of the classroom, then seeing *letters and words, numbers and edges,* the teacher's face, smiles of classmates).

5. For another poem about seeing sharply (through a microscope), look for **"Armor" by Margarita Engle** (3rd Grade, Week 8, page 156). Or look for the fun pop-up picture storybook *Arlo Needs Glasses* by Barney Saltzberg.

## WEEK 26: KITCHEN SCIENCE

### FIRST SCIENCE PROJECT
by **Lesléa Newman**

I ate the avocado
That we bought at the store,
And it was so delicious
I wished that I had more.

My mother cleaned the pit up,
And handed it to me.
"Get some toothpicks and we'll grow
An avocado tree."

We pierced the pit with toothpicks
And perched it on a glass,
Then filled the glass with water.
"Now watch what comes to pass."

Soon some roots were dangling down,
Straggly, thin, and white.
Soon we saw a bright green stem
That reached up toward the light.

Soon the plant was growing leaves,
We welcomed every one.
Each leaf so green and shiny,
Unfurling towards the sun.

And when the plant grew bigger,
We planted it in dirt.
The next day it was taller,
An overnight growth spurt!

The plant grew even bigger,
Three feet from stem to root.
And though we watched and waited,
It never did bear fruit.

"It doesn't really matter,"
My mother said to me.
"We still have something lovely:
An avocado tree!"

### Take 5!

1. If you can plan ahead, **bring an avocado or avocado pit as your poetry prop.** Read the poem aloud and pass the avocado or pit around for students to examine. Or for visuals and more, go to AvocadoCentral.com.

2. Read the poem aloud again and **invite students to say the important last line together**.

3. **Use this poem to talk about the parts of the avocado plant**—fruit, peel, pit, roots, stem, leaves, tree. Then watch a time lapse video of an avocado pit growing (e.g., YouTube.com/watch?v=KM6muAZphdY)

4. Each stanza of the poem reports another stage in the avocado investigation. **Challenge the students to work in pairs or trios to draw a picture for one stanza,** then post all the pictures in order corresponding to the poem.

5. For a poem about a classroom plant, look for **"The Class Plant" by Janet Wong** (2nd Grade, Week 5, page 113), or find the informational picture book *Experiments with Plants* by Christine Taylor-Butler.

## Week 27: Video Technology

**First Grade**

### Take 5!

1. If possible, **play a few moments from a television or Internet news channel without sound to set the stage.** Then read the poem aloud, pausing before the final line.

2. Both stanzas end with the word *together*, so **invite students to chime in on the final word as you read the whole poem aloud again.** Add a gesture, like hands clasping, for emphasis.

3. For discussion: *What makes a person good at puzzles?* (Patience, careful observation, noting details.)

4. The poem highlights skill with video tools. **Collaborate with students to create a quick digital collage** about the class, school, or community (e.g., using Glogster.com). Show the students the choices of text, fonts, color, graphics, and even animation, if possible, while you input those items and create the finished product.

5. For another poem about a mother who uses science at work, look for **"Geologist" by Betsy Franco** (1st Grade, Week 34, page 102).

### Pieces
by **Renée M. LaTulippe**

My Mamma edits videos
for nightly news and TV shows.
She knows where all the pieces go
and puzzles them together.

She often lets me see the screen
and watch the frames go, scene by scene.
A piece of person, sky, and tree:
we puzzle it together.

## WEEK 28: MACHINES
### FIRST GRADE

### LEVERS
by **Michael Salinger**

A screwdriver
opening a big can
of house paint
is a machine
that is simple
and clever.

Just a beam
and a fulcrum
distributing force
and you've made yourself
a lever.

### Take 5!

1. After reading this poem aloud, **do a quick demonstration of a lever in action** (use a screwdriver, pencil, ruler, or even your arm and elbow). Talk about what a "lever" is—something that helps us move things more easily by magnifying the force or effort. One example from the Web is YouTube.com/watch?v=QP1U2d7_GLU.

2. Read the poem aloud again and **invite students to say the important last line together** with you.

3. **Look around the room for any other examples of levers** (door handles, cabinet handles, scissors, tweezers, seesaw, etc.) and talk with students about how they are useful in everyday life.

4. **Work with students to investigate how simple machines use movement** like levers and other gears to function. Consider electrical switches, doorknobs, pencil sharpeners, etc. Discuss how objects can move in a straight line, zig zag, up and down, back and forth, round and round, and fast and slow.

5. Pair this poem with another about simple machines by the same poet, **"Gears" by Michael Salinger** (2nd Grade, Week 28, page 136), and with a poem about building things, **"The Crane Operator" by Rebecca Kai Dotlich** (3rd Grade, Week 29, page 177) or Dotlich's picture book *What Can a Crane Pick Up?*

## WEEK 29: BUILDING THINGS
### FIRST GRADE

**Take 5!**

1. Before reading this poem aloud, **it may be helpful to show what driftwood looks like.** If you don't have access to an actual piece, look for images online.

2. This poem is also full of numbers, so **invite students to chime in on all the number words** while you read the whole poem aloud again. Cue them with note cards with the numbers on them: *two, thirty, twenty, three.*

3. Encourage students to **share experiences with building things on their own** (or with help) with sticks, blocks, LEGOS, boxes, etc.

4. Talk with students about **how we can use natural resources and materials around us to create something new.** Use this poem to discuss the necessary steps (e.g., find a partner, gather and organize materials, estimate how much you'll need, start building, adjust for mistakes and improvements, enjoy your product).

5. For another poem about scrounging for old materials to make something new, share **"Tinker Time" by Janet Wong** (Kindergarten, Week 29, page 57). Also look for more poems about building with *Dreaming Up: A Celebration of Building* by Christy Hale, or for a playful approach to building, look for *The LEGO Book* by Daniel Lipkowitz.

## DRIFTWOOD HUT
by **Renée M. LaTulippe**

Today we'll build a driftwood hut,
my brother and I, we two.
A high-tide haul of branches means
that we have work to do!

We stack up thirty sea-slick sticks,
and figure that it's plenty.
But stacking sure is fun, and so
we stack an extra twenty.

The trick is lashing them. You need:
some twine, some time, a brother.
Two go up. Three fall down—
oops! Stack and lash another.

We wrap the hut with tattered tarp
and seaweed washed ashore,
then gather dune-grass, twigs, and shells
to pave our sandy floor.

Our driftwood hut is secret—thick
with salt-wind rushing through.
Just big enough to whisper in—
my brother and I, we two.

## WEEK 30: SCIENCE FAIR

**FIRST GRADE**

### MY PROJECT FOR THE SCIENCE FAIR
by **Kenn Nesbitt**

My project for the Science Fair
was positively cool.
I built myself a time machine
and showed it off at school.

Inventing it was not too hard;
I had a little aid.
My future self came back in time
and showed me how they're made.

### Take 5!

1. Before sharing this poem, **post two identical photos of yourself side by side as your poetry prop.** Then read the poem aloud, with particular emphasis on the last two lines.

2. Share this poem again, and **invite students to read the important third line of each stanza** (*I built myself a time machine; My future self came back in time*) while you read the rest of the poem aloud.

3. **Brainstorm possible science fair project ideas with students.** Consult this video series recommended by the National Science Teachers Association: JPL.NASA.gov/education/sciencefair/.

4. **Talk with students about what a scientist is and explore what different scientists do.** Here's one excellent site with multiple links: ScienceBuddies.org/science-fair-projects/science_careers.shtml.

5. Pair this poem with another about entering the science fair, **"Science Fair Day" by Eric Ode** (Kindergarten, Week 30, page 58).

## Week 31: Famous Scientists

**First Grade**

### Take 5!

1. Read this poem aloud, pausing between stanzas for added effect. Then **show pictures of da Vinci and some of his drawings and inventions** found at DrawingsofLeonardo.org.

2. Share the poem again, and **invite students to read the first line of each stanza** *(In Italy, long, long ago / He was... / who dreamed up... / In fact... )* while you read the rest of the poem aloud.

3. Da Vinci was both an artist and a scientist. **Talk about how drawing and inventing might go together.**

4. Use the examples in this poem to talk with students about how **information and critical thinking are used in scientific problem solving** in Leonardo's time (1500s) and now. What kinds of inventions might they imagine for the future? For background information, go to Legacy.MOS.org/sln/Leonardo/.

5. For a poem about inventing things in the future, look for **"Invention Intentions" by Kristy Dempsey** (3rd Grade, Week 34, page 182).

## Da Vinci Did It!
### by Renée M. LaTulippe

In Italy, long, long ago,
a genius lived—
LEONARDO!

He was—
a painter, sculptor, mathematician,
engineer, and skilled musician

who dreamed up—
robots, carts, and parachutes,
flying planes and diving suits.

In fact—
as long as time did not forbid it,
you can bet da Vinci did it!

## WEEK 32: MORE FAMOUS SCIENTISTS

### THE "BLACK LEONARDO"*
by **J. Patrick Lewis**

*George Washington Carver*
*1846–1943*
*Botanist, educator, soil scientist, and inventor*


He analyzed the peanut
   And the sweet potato too,
Developing things like plastics, paints,
   Linoleum, shampoo,
Peanut butter, vinegar,
   Insecticide, and yeast.
"Sometimes," he said, "you find the
   Secrets most among the least."


*\*Time Magazine reference from an article published on November 24, 1941*

### Take 5!

1. If possible, **find one or more of the objects listed in poem to feature as a poetry prop** (*peanut, sweet potato,* something *plastic, paint, linoleum, shampoo, peanut butter, vinegar, insecticide, yeast*). Then read the poem aloud slowly.

2. For a repeated reading, **invite students to chime in on their favorite Carver invention** (*plastics, paints, linoleum, shampoo, peanut butter, vinegar, insecticide, yeast*) while you read the rest of the poem aloud.

3. **Talk about why the poet titled this poem "The 'Black Leonardo.'"** (Leonardo da Vinci was a famous artist and inventor, and George Washington Carver was a famous scientist and inventor who was African American.)

4. Work with students to use this poem and web resources to **make a list of some of the products Carver discovered could come from peanuts and sweet potatoes** (starting with *plastics, paints, linoleum, shampoo, peanut butter, vinegar, insecticide, yeast*). Look for more details about Carver and his work in *The Little Plant Doctor* by Jean Marzollo.

5. Make the connection to Leonardo Da Vinci explicit by revisiting the poem **"Da Vinci Did It!" by Renée M. LaTulippe** (1st Grade, Week 31, page 99).

## WEEK 33: COMPUTERS

**FIRST GRADE**

### Take 5!

1. Before sharing the poem, **show an image of a bridge in the background.** Choose one from HistoryofBridges.com, for example. Then read this poem aloud with enthusiasm.

2. Share the poem again, but this time **invite students to read the crucial middle stanza** *(I use computers. I use my brain. / I think and test till the answer is plain.)* while you read the rest of the poem aloud.

3. **Collaborate with students to create a quick glog,** a digital, interactive poster (using Glogster.com), pulling together images for key words from the poem in a new, visual representation of the poem's topic. Show the students the choices of text, fonts, and images, while you input those items and create the finished product.

4. **Use this poem to help students describe what engineers do.** Make a list of all the inventions included in the poem (bridge, wheel, robot, rocket, electronic device, running shoes, anti-snoring device). Talk about the tools needed to create these inventions (computers, brain, tests).

5. In contrast, share a poem that raises questions that a scientist might tackle with **"Late Night Science Questions" by Greg Pincus** (2nd Grade, Week 3, page 111). And look for David Macaulay's "readers" that explain how things work, like *Toilet: How It Works*.

## THE ENGINEER
### by **Stephanie Calmenson**

Listen up and you will hear
Why I am called an engineer.

I solve. I build. I invent.
I'd say my time is very well spent.

Want a bridge? I'll design it for you.
Want a new kind of wheel? I'll develop that, too.

I use computers. I use my brain.
I think and test till the answer is plain.

Want a robot, a rocket, an electronic device?
I'll take the assignment. I won't think twice.

I'll make running shoes that will send you soaring!
I'll develop a device that will keep you from snoring!

My life is all about invention.
Making the world work better is my intention.

*The Poetry Friday Anthology for Science*

# WEEK 34: SCIENCE CAREERS

**FIRST GRADE**

## GEOLOGIST
by **Betsy Franco**

My mother and I are
out in the field again
at Medicine Lake Volcano
where bald eagles
own the sky.

Adding hammer sounds
to the semi-silence,
we break rocks—
some filled with white flecks
some with yellow—
to unlock the story of
the lava's flow.

Walking up a cinder cone
our feet sink into
fine sand.
Below us lies
the dry lava bed
we map in
four dimensions:
width, length, depth,
and time.

In a soft voice
my mother says
she knows
this place better
than anyone.

### Take 5!

1. **Set the stage by sharing a video of the location** mentioned in the poem, Medicine Lake Volcano. Search for your favorite image at Flickr.com. Then read the poem aloud slowly and quietly.

2. Read the poem again and this time **invite students to add movements** while you read aloud—arms extended like soaring eagles for the first stanza, hammering (not too loudly) during the second stanza, and walking in place during the third stanza.

3. Being a geologist is one career possibility in science. **Talk about some other science careers** that might interest students. For a wide variety of real life examples, go to SmithsonianEducation.org/Scientist/.

4. Use the details in this poem to talk about how **the natural world includes rocks, soil, and water** that can be observed and studied in many ways (e.g., *we break rocks / some filled with white flecks / some with yellow / to unlock the story of / the lava's flow; cinder cone, fine sand, dry lava bed.*) If possible, bring in a sampling of soil or rocks and guide students in describing all their properties.

5. For another volcano poem, look for **"Science Project" by Lee Wardlaw** (2nd Grade, Week 30, page 138). Also look for Seymour Simon's photo-essay picture book *Volcanoes* for more images and information.

# Week 35: Future Challenges

**First Grade**

## I Want to Know Why
by **David L. Harrison**

I want to know why
People get sick.
I want to know why
Little kids die.
I want to help find
Cures for diseases.
Someone must do it.
Why not I?

I want to know why
People are starving,
Why kids suffer
Such misery.
I want to help find
Cures for famine.
Someone must do it.
Let it be me.

### Take 5!

1. Before reading this poem aloud in a soft voice, point out to students that **many poems are funny, but some are serious**—like this one.

2. Read the poem aloud again, and **invite students to chime in when the title of the poem appears within the poem** (*I want to know why*). Cue them by holding up a card with the phrase written on it.

3. **Talk about opportunities that students have for helping locally and globally.** Research local food pantries, soup kitchens, and volunteer projects at area hospitals.

4. **Talk about how scientists try to identify and explain a problem and then find solutions.** Discuss possible questions students have about how to make the world a better place.

5. For another serious poem about a problem scientists are tackling, look for **"Cancer" by Mary Lee Hahn** (3rd Grade, Week 35, page 183).

## WEEK 36: FUTURE DREAMS

**FIRST GRADE**

### EVERYDAY ASTRONAUT
by **Carmen Tafolla**

When they said that they were looking
for everyday astronauts
people from just normal life
Teachers with totes filled with things to surprise
Bus drivers with steady and focused dark eyes
Students peeking out from under stacks and reams . . .

I started to dream.
To see myself floating in space
waving at friends from a moon away
a planet my nightlight, shining soft, night and day

Our little blue planet is a warm, cozy room
in this wide and wonderful
wild and wakened
universe
we call home


### UN ASTRONAUTA COMÚN
por **Carmen Tafolla**

Cuando oí que buscaban
astronautas ordinarios
entre la genta de vida sencilla . . .
Imaginé maestros con mochilas llenas de sorpresas,
Choferes de camión con ojos de aguilón,
Jóvenes muy normales con sus juegos digitales

Mi mente se subió
a una astronave suave
Mis fantasías y sueños
comenzaron a volar
Ahora floto en el espacio
saludando amigos desde una luna lejana

Mi lucecita nocturna es un planeta azul
que constante brilla, de noche y de día,
Mi habitación pequeña es el sistema solar,
un rinconcito cómodo de esta casa grande,
enorme universo,
que llamamos hogar.

### Take 5!

1. It's time to pretend! Read the poem aloud, and as you get to the second stanza, slow down and **wander around the room as if floating through space,** ending back "home" in the final line.

2. Since this is a bilingual poem, it's the perfect opportunity to **invite someone who is fluent in Spanish (in your class, school, or community) to join you by reading the Spanish poem** before or after you read the poem in English. If necessary, you could use VoiceThread to make a long distance recording in advance of a Spanish Speaker reading the Spanish poem.

3. For discussion: ***What might it be like to be an astronaut someday?*** Look at SpaceCampage.com for details and photos.

4. Although this poem IMAGINES floating in space, **talk about what an astronaut would likely see in space, beginning with examples from this poem** (moon, planets). NASA.gov is full of images and resources to share.

5. Revisit a previous poem that imagines what it would be like to visit Venus, **"Did You Know?" by Julie Larios** (Kindergarten, Week 14, page 42). Also look for *Astronaut Handbook* by Meghan McCarthy for more about what it takes to become an astronaut and what astronauts do.

# Poems for Second Grade

# NGSS Science and Engineering Practices: Second Grade

*These practices form the foundation of disciplinary literacy in science and integrate reading, writing, listening, and speaking skills from the language arts. Here we indicate which weekly poems emphasize which science and engineering practices at each grade level.*

| PRACTICE | POEM |
|---|---|
| Asking questions and defining problems | Weeks 3, 28, 32 |
| Developing and using models | Weeks 12, 19, 29, 30 |
| Planning and carrying out investigations | Weeks 4, 34 |
| Analyzing and interpreting data | Weeks 14, 15, 18, 20, 22, 24 |
| Using mathematics and computational thinking | Weeks 7, 8, 17 |
| Constructing explanations and designing solutions | Weeks 5, 6, 27, 35, 36 |
| Engaging in argument from evidence | Weeks 2, 11, 25, 26 |
| Obtaining, evaluating, and communicating information | Weeks 1, 9, 10, 13, 16, 21, 23, 31, 33 |

# Second Grade

| | | |
|---|---|---|
| week 1 | Scientific Practices | Let's All Be Scientists! *by Renée M. LaTulippe* |
| week 2 | Lab Safety | Pass Me Those Ear Muffs *by Graham Denton* |
| week 3 | Ask and Ask Again | Late Night Science Questions *by Greg Pincus* |
| week 4 | Observations | Discovery/Descubrimiento *by Margarita Engle* |
| week 5 | Predictions & Hypotheses | The Class Plant *by Janet Wong* |
| week 6 | Investigations | My Experiment *by Julie Larios* |
| week 7 | Data | Crazy Data Day *by Janet Wong* |
| week 8 | Tools of Science | Meter Stick *by Amy Ludwig VanDerwater* |
| week 9 | Matter | Imagine Small *by Eileen Spinelli* |
| week 10 | More Matter | Liquids Can't Contain Themselves *by Heidi Bee Roemer* |
| week 11 | Force, Motion & Energy | Gravity *by Joyce Sidman* |
| week 12 | More FM&E | Go Fly a Kite *by Laura Purdie Salas* |
| week 13 | Light & Sound | Sound Waves at Breakfast *by Susan Marie Swanson* |
| week 14 | Space | Uh Oh, Pluto *by Jeannine Atkins* |
| week 15 | Sun, Earth & Moon | Earth's Tilt *by Douglas Florian* |
| week 16 | The Water Cycle | Water Round *by Leslie Bulion* |
| week 17 | Weather & Climate | Rain Gauge *by Anastasia Suen* |
| week 18 | Forces of Nature | Plates *by Ann Whitford Paul* |
| week 19 | Soil & Land | Glacier *by Kate Coombs* |
| week 20 | Natural Resources | Fossil Fuels *by Janet Wong* |
| week 21 | Ecosystems | The Rain Forest *by Bobbi Katz* |
| week 22 | Adaptations & Traits | The Leopard Cannot Change His Spots *by Lesléa Newman* |
| week 23 | Cycles | Becoming Butterflies *by Jeannine Atkins* |
| week 24 | Patterns | Rings Not Letters *by Juanita Havill* |
| week 25 | Human Body | Shots! Shots! Shots! *by Joy Acey* |
| week 26 | Kitchen Science | Mold *by Charles Waters* |
| week 27 | Video Technology | Game Programmer *by Janet Wong* |
| week 28 | Machines | Gears *by Michael Salinger* |
| week 29 | Building Things | My Robot *by David L. Harrison* |
| week 30 | Science Fair | Science Project *by Lee Wardlaw* |
| week 31 | Famous Scientists | Jane Goodall Begins a Speech *by Susan Marie Swanson* |
| week 32 | More Famous Scientists | Wondering Why *by Shirley Smith Duke* |
| week 33 | Computers | (Super)Power: (to the)Point *by Kristy Dempsey* |
| week 34 | Science Careers | Water Engineered *by Sara Holbrook* |
| week 35 | Future Challenges | The Lament of Lonesome George *by Jane Yolen* |
| week 36 | Future Dreams | Space Yacht *by Juanita Havill* |

"If I had to live my life again, I would have made a rule to **read some poetry and listen to some music at least once every week**."

≈ Charles Darwin ≈

## Second Grade

### Week 1: Scientific Practices

### Take 5!

1. **Draw a question mark on the board** and talk about how everybody likes to ask questions, particularly scientists and poets who like to wonder about things. Then read this poem aloud, pausing briefly at the end of each line.

2. Each line begins with the same word, *Let's.* **Invite students to chime in on that repeated word (*Let's*)** as you read the whole poem aloud again. Pump your fist in the air as a cue for students to join in.

3. For discussion: ***Which is more important: questions or answers?*** Talk about how BOTH are important.

4. Invite students to **identify what is involved in problem solving based on the steps presented in this poem**: choosing a subject, asking questions, guessing, planning, experimenting and investigating, watching, writing, learning, and showing or sharing. Talk about how each of these is an important part of solving problems.

5. Share another poem about asking questions, **"Backwards" by Janet Wong** (1st Grade, Week 3, page 71), or connect with a simple introductory nonfiction book like *What Is a Scientist?* by Barbara Lehn or *You Are a Scientist* by Marcia S. Freeman.

### Let's All Be Scientists!
by **Renée M. LaTulippe**

Let's choose a subject that we love.
Let's ask a thoughtful question.
Let's guess at what the answer is.
Let's plan the investigation.

Let's make a great experiment.
Let's watch and write and learn.
Let's see—did we find answers? Yes!
Let's show what we know to the world!

## WEEK 2: LAB SAFETY

### PASS ME THOSE EAR MUFFS
by **Graham Denton**

In our
science lesson
on sound,
Miss Butters asked us
to name
"one situation
where ear protection
might have to be worn
to save hearing
from being
damaged."

I didn't write
when working
with power tools
or with noisy
machinery
or even
at a
very loud
rock
concert;

I wrote,
"In the classroom,
listening to
Miss Butters shout
when she's really,
really
mad at me
for being so cheeky."

And do you
know what,
after Miss Butters
read that,
I was proved
absolutely
right.

### Take 5!

1. If you have ear protection handy, **don a set of earphones, earbuds, or earmuffs as a poetry prop** while you read this poem aloud. You may need to explain the word *cheeky* (British for *smart-alecky*).

2. Next, read the poem aloud again, this time having **students read the lines "written" by the child in the poem** (*In the classroom, / listening to / Miss Butters shout / when she's really, / really / mad at me / for being so cheeky*) while you read the rest of the poem.

3. Use the details in this poem to **talk about when it might be important to protect your ears** (around power tools or noisy machinery, at a loud concert, in extreme cold or wind, from the sun, etc.).

4. Ear protection is one important component of lab safety. **Talk about other important safe practices necessary for classroom and outdoor investigations,** including wearing safety goggles, washing hands, and using materials appropriately. Look for help from *Hands-On Science: Sound and Light* by Jack Challoner and Maggie Hewson.

5. For more poems about sounds, look for **"Listen" by Amy Ludwig VanDerwater** (Kindergarten, Week 13, page 41) and the book *A Rumpus of Rhymes: A Book of Noisy Poems* by Bobbi Katz.

## WEEK 3: ASK AND ASK AGAIN

**SECOND GRADE**

### LATE NIGHT SCIENCE QUESTIONS
by **Greg Pincus**

Do sneakers make me fast?
How long does winter last?
Is goop the same as goo?
What can't a robot do?
What makes a motor go?
Can we drink $H_3O$?
Why is the ocean deep?
Why do I have to sleep?

### Take 5!

1. If you have a cell phone with a "Siri" kind of function, **try asking it one of the questions in this poem just for fun before reading this poem aloud**. Or simply draw a giant question mark on the board before launching into the poem.

2. Invite students to select their favorite question line and to **chime in on that line only** while you read the rest of the poem aloud.

3. Challenge students to work in pairs or trios to **choose one of the poem questions and quickly research possible answers.** Then come together to share information.

4. Use this poem to talk with students about **how scientists ask questions as part of their scientific inquiry and investigations**. This includes questions about organisms (e.g., *Is goop the same as goo?*), objects (e.g., *What can't a robot do?*), and events (e.g., *How long does winter last?*).

5. For another poem about the multitude of possibilities in science, look for **"The Engineer" by Stephanie Calmenson** (1st Grade, Week 33, page 101). Also seek out the question-based poetry book *Where Fish Go in Winter and Answers to Other Great Mysteries* by Amy Goldman Koss.

## Week 4: Observations

**Second Grade**

### Discovery/Descubrimiento
by/por **Margarita Engle**

Silence.  
Careful.  

We can't be noisy  
if we want to see  
shy birds.  

I never knew  
that explorers  
have to be  
so patient.  

Look!  
Such a beautiful  
hummingbird!  

The wait  
is worth it!  

Silencio.  
Cuidado.  

No podemos hacer ruido  
si queremos ver  
pájaros tímidos.  

Yo no sabía  
que los exploradores  
tienen que ser  
tan pacientes.  

¡Mira!  
¡Que colibrí  
tan lindo!  

¡Vale la pena  
esperar!  

### Take 5!

1. Before reading this poem, **play a video clip** from AllAboutBirds.org to set the stage or as a background while you read the poem aloud.

2. Read the poem aloud again and **invite students to say the first two and last two lines with you** to "bookend" the poem.

3. Since this poem is presented in Spanish and English, **invite a reader and speaker of Spanish on your campus or in your community to read the Spanish version aloud** while you (or another volunteer) read(s) the English version aloud again. Record both readings for future reference, if possible.

4. The poet provides several details about the patience needed to observe birds outdoors. If possible, take a moment for bird watching with students (outside or from a window) and **work together to model note taking and observation**.

5. For another poem about what we can observe outdoors, look for **"Step Outside. What Do You See?" by Allan Wolf** (Kindergarten, Week 4, page 32) **or "Classroom in the Meadow" by Jeannine Atkins** (3rd Grade, Week 4, page 152). For a nonfiction resource, consult *About Birds: A Guide for Children* by Cathryn Sill and John Sill, and for more nature poems look for *Nest, Nook & Cranny* by Susan Blackaby.

*The Poetry Friday Anthology for Science*

## WEEK 5: PREDICTIONS & HYPOTHESES

**SECOND GRADE**

### THE CLASS PLANT
by **Janet Wong**

The plant on our bookshelf
is turning yellow, drooping
and dropping leaves.

We talk about what it might need.
I feel like a doctor in a hospital
talking about a sick patient.

Jack wants to give it more water.
Kayla says the soil smells moldy
and feels soggy—too much water.

Chris asks: How about more light?
We move the plant to a bigger pot.
I place it carefully near the window.

I have a prediction: "Tomorrow
our plant will be dancing!"
They laugh but they will see.

The next day the weather is very hot.
The air conditioning is blowing hard
right down on the leaves of our plant.

    Hot weather
+  air conditioning
    Dancing Plant!

### Take 5!

1. Add a bit of fun to sharing this poem with a poetry prop—**show a (drooping) plant before reading the poem aloud.**

2. For a poem like this with multiple characters, **create an impromptu readers theater performance of the poem.** Ask for student volunteers to read the lines about Jack (*Jack wants to give it more water*), Kayla (*Kayla says the soil smells moldy / and feels soggy—too much water*), Chris, (*Chris asks: How about more light?*), and the poem's speaker (*Tomorrow / our plant will be dancing!*). You read the rest of the poem aloud.

3. For discussion: **Will this plant survive?** What should they do next for the plant?

4. The poem presents a problem with several hypotheses about what is going wrong and what is needed next. **Talk about how scientific problem solving includes identifying and explaining a problem, proposing a solution, making predictions, and observing details and outcomes.** Pinpoint examples of these steps from the poem.

5. For another plant project, pair this poem with **"First Science Project" by Lesléa Newman** (1st Grade, Week 26, page 94). Or look for **"Testing My Hypothesis" by Leslie Bulion** (3rd Grade, Week 5, page 153) for another science connection.

## WEEK 6: INVESTIGATIONS

**SECOND GRADE**

### MY EXPERIMENT
by **Julie Larios**

I tried each possibility,
I tried it all, I tried my best,
I tried to think, I tried to see,
I tried things out, I didn't rest,
I thought I had it, I thought I knew,
I thought what I had was good and true,
but the bottom caved in, the top spilled out,
I couldn't figure the darn thing out,
it all collapsed, it all fell down,
the smile I smiled became a frown.
I didn't succeed, so tomorrow is when
I have to try and try again.
That's good advice, that's right, I guess—
but meanwhile (*sigh) what an awful mess.

### Take 5!

1. Before sharing this poem, take a moment to **encourage students to think about a time when something did NOT work out, when they tried something but failed.** This poet understands that. Then read the poem aloud.

2. In the next reading, **coach students to say the repeated phrase *I tried* each time it appears in the poem** while you read the rest aloud. Cue them by holding up a card with *I tried* written on it.

3. For discussion: *Failure may be even more important than success. Why?*

4. **Talk about failure and how important it is in science and in life**, particularly if we reflect on why we failed and how to adjust or improve next time. Remind students that scientists often repeat experiments many times as they search for answers and discoveries, and that other scientists must also be able to repeat the experiment for it to be considered valid.

5. For another poem about trial and error, look for **"Paper Airplanes" by Janet Wong** (5th Grade, Week 3, page 231). Pair this poem with the picture book *11 Experiments that Failed* by Jenny Offill and Nancy Carpenter.

## WEEK 7: DATA

**SECOND GRADE**

### Take 5!

1. If possible, **have a toy car prop ready as you read this poem aloud**. Scoot it along a surface for effect throughout the reading.

2. For a follow up reading, **invite students to say the number words** in the poem (*four, six, nine, three*) while you read the rest of the poem aloud. Cue them by holding up the corresponding number of fingers.

3. **Talk with students about the data in the poem** and make a simple chart together to show the results in this poem experiment.

4. **Use this poem to point out that scientists use clocks, timers, and stopwatches** to collect and record information. Try using a clock or stopwatch for an on-the-spot investigation comparing how far various objects (toy car, cup, pencil) might roll on a flat or inclined surface in six seconds. Try one of the same items several times, time it each time, and discuss the results. Did it change? Why or why not?

5. Link this poem with another poem about measurement by **Janet Wong**, **"Stopwatch"** (Kindergarten, Week 8, page 36), or with another poem about testing a car's speed, **"Designing an Experiment" by Avis Harley** (5th Grade, Week 6, page 234).

## CRAZY DATA DAY
### by Janet Wong

We rolled a car down the ramp
four times
and timed it with my stopwatch.

Six seconds.
Six seconds.
Six.
And six.

And then
the car
started playing tricks!

The next time was
NINE!

I cleaned the wheels.
George checked the track.
We rolled again.
Did the six come back?

What?
THREE!
No way!

What a Crazy Data Day!

## WEEK 8: TOOLS OF SCIENCE

SECOND GRADE

### METER STICK
by **Amy Ludwig VanDerwater**

It's a pleasure to measure in meters.
It's a pleasure to measure because
everything measures in units of ten.
It measures so sweetly.
It does.
Ten centimeters are one decimeter.
Ten decimeters, a meter.
Divide.
Multiply.
Always by ten.
Measuring couldn't be neater.
And when I must measure
a plant or a pencil
when I must measure
a scrap of my day
I am connected
to all those who measure
in meters
in countries
so far
        far
                away.

### Take 5!

1. **What is the ideal prop for this poem? A meter stick or ruler showing metric measurement.** (Find various printable paper rulers at Vendian.org/mncharity/dir3/paper_rulers/.) Show the prop and then read the poem aloud slowly and clearly.

2. In sharing the poem aloud again, **students can say the key word, *ten*, each time it occurs in the poem.** Cue students by raising both hands, signaling the number 10.

3. Research together a list of countries that use the metric system for measurement, and **talk about how this is one of many tools scientists use** around the world.

4. Measurement is an important component in scientific investigation. It's part of collecting, recording, and comparing information. **Work with students to identify a tool that can approximate a metric unit** (a strip of paper, a length of string) and then use it to measure and compare several everyday objects that are handy. Look here for a set of printable, cut-out metric rulers: Farr-integratingit.net/Integration/General/Metric/rulers_cm_in_simple.pdf.

5. Match this poem with another on metrics, **"Celsius Thermometer" by Renée M. LaTulippe** (1st Grade, Week 8, page 76), or with another about measuring, **"My Bean Plant" by Amy Ludwig VanDerwater** (Kindergarten, Week 7, page 35).

## WEEK 9: MATTER

### SECOND GRADE

### Take 5!

1. Before sharing this poem, **invite students to close their eyes and imagine the smallest thing they can think of.** Then read this poem aloud, pausing before each stanza for added effect.

2. Then read the poem aloud again and **invite students to chime in on their favorite small item in the poem** (*a rubber duck, a ping-pong ball, pebble, violet, baby's tooth, pumpkin seed, freckle, flea, gnat, a speck of dander from the cat, a teeny-tiny molecule, the atom*) while you read the rest of the poem aloud.

3. Collaborate with students to **create a quick glog, a digital interactive poster (using Glogster.com), pulling together images and key words from the poem** in a new, visual representation of the poem's theme. Show the students the choices of text, fonts, color, graphics, and even animation, if possible, while you input those items and create the finished product.

4. **Use this poem to talk about the physical properties of objects.** Make a list of the small objects in the poem (*a rubber duck, a ping-pong ball,* and so on) and identify the shape and relative size of each. What is the pattern in their sequencing in the poem? (They get smaller and smaller.)

5. For another poem about invisible particles, check out **"Think of an Atom" by Buffy Silverman** (5th Grade, Week 9, page 237). Also look for a helpful nonfiction picture book like *Atoms and Molecules* by Molly Aloian.

## IMAGINE SMALL
### by **Eileen Spinelli**

Imagine something very small:
a rubber duck, a ping-pong ball.

Imagine something smaller yet:
a pebble or a violet.

Go smaller now: a silver bead,
a baby's tooth, a pumpkin seed.

Keep going—
freckle, flea, or gnat,
a speck of dander from the cat.
Imagine that.

And then imagine this—so cool!—
a teeny-tiny molecule.
So teeny-tiny you and I
can't see it with the naked eye.

To think of it gives me a chill.
But there is something smaller still:
the atom!

Billions fit in a fleck of foam
or on the dot at the end of this poem.

Billions.

*The Poetry Friday Anthology for Science*

## LIQUIDS CAN'T CONTAIN THEMSELVES
### by **Heidi Bee Roemer**

Sticky honey leaks from a jar.
Oozy ketchup squirts too far.
Hot soup overfills its bowl;
Liquids dribble and ripple and roll!

A bucket, a jug, a jar, or a vase
keeps each liquid in its place.
But liquids cannot keep a shape;
they're always seeking to escape.

Oops! You spilled your carton of juice.
Aren't liquids messy when they get loose?

### Take 5!

1. For this poem, **the obvious poetry prop is a container (a mug, cup, or jar) full of water**, plus some paper towels for spills. Read the poem aloud and if you're feeling brave, make a few small spills, too.

2. Sometimes poets use words that sound like sounds, which is called *onomatopoeia*. **Read the poem aloud again with students chiming in to highlight the sound words** *(squirts, dribble, ripple)*.

3. Use a cup or mug and a flat bowl or plate and put the same amount of water in both. **Talk with students about how these hold the same amount of water,** despite the differing measurement tools we're using. Use a measuring cup to show children the equal amounts before pouring (or after).

4. **Make a list of the liquids mentioned in the poem and arrange them on a continuum** from the runniest to the thickest liquid state (juice, soup, ketchup, honey).

5. For another poem about measuring liquids, look for **"Thirsty Measures" also by Heidi Bee Roemer** (5th Grade, Week 26, page 254).

## Week 11: Force, Motion & Energy

**Second Grade**

### Take 5!

1. Use web resources like Google.com/earth to **view images of earth from space to set the stage** before reading this poem aloud.

2. Read the poem aloud again and **invite students to say the last two lines together** with you.

3. For discussion: *If the moon is earth's sister, how do the other planets fit into the "family"?*

4. **Use this poem to talk about the concept of gravity** (on a global scale) and how things on earth (like rocks, trees, elephants, clouds, kites, air) can move, but are bound back to the earth by the "hug" of gravity. If possible, try some quick experiments with throwing or dropping objects to demonstrate.

5. Connect this poem with another about gravity, **"Thank You, Isaac Newton" by Eileen Spinelli** (Kindergarten, Week 12, page 40).

## Gravity
### by Joyce Sidman

Think of the Earth
as a mama
with a warm, heavy heart.
She's lonely in space.
She reaches out her great arms
and holds us to her:
rocks, trees, elephants,
clouds, kites, air.
We can fly away—of course!
But only so far
before she calls us back.
We can jump
and vault and bounce and twirl;
but always, always,
we return to her.
She worries about growing
older, smaller,
weaker—
like her bleak sister,
the moon.

She holds on tight.
Her hug
encircles the world.

*The Poetry Friday Anthology for Science*

## SECOND GRADE
### WEEK 12: MORE FORCE, MOTION & ENERGY

**GO FLY A KITE**
by **Laura Purdie Salas**

Above the kite, the pressure's low.
The air's a streaming, breezy flow.

Below the kite, the pressure's higher.
Up! Up! Up! This one's a fly-er!

Lift versus drag.
Lift wins!
That's why . . .

your kite
breaks
free
and
climbs
the
sky!

### Take 5!

1. **Before sharing this poem, share images of kites flying** (e.g., from IntoTheWind.com or Kites.com). Then read the poem aloud, pausing dramatically between stanzas.

2. In sharing the poem aloud again, **students can say the repeated word *Up! Up! Up!* in the poem.** Cue students by pointing up higher, higher, and higher.

3. **Talk about other inanimate objects that can "fly"** (like Frisbees, paper airplanes, baseballs, jets, helicopters, etc.).

4. Use the details in this poem to **talk about factors that influence the kite in flight**: pressure, streaming, breezes, lift, and drag. Check out NASA's "Beginner's Guide to Aerodynamics" for help, available at GRC.NASA.gov/WWW/K-12/airplane/bga.html.

5. Pair this poem with another about a flying object, **"Frisbee" by Glenn Schroeder** (1st Grade, Week 12, page 80), and don't miss the poems in *Give Me Wings* edited by Lee Bennett Hopkins.

## WEEK 13: LIGHT & SOUND

**SECOND GRADE**

## SOUND WAVES AT BREAKFAST
### by Susan Marie Swanson

I can hear the garbage truck backing up
*beep beep beep*,
and then the big roar and ruckus of its work.
My neighbor's dog goes crazy *bark bark barking*.
My *clang* baby sister *clang*
bangs her spoon *clang* on her tray.

*Shhh.*

I'm listening for the vibrations of squirrel claws
scritching up trees.

I like the sound waves from the dog's clinking tags
and his tail thumping on the wooden stoop.

My ears are ready for the pop
when the lid pops off this new jar of jam.

### Take 5!

1. **To set the stage for this poem, play a video of a noisy garbage truck** (e.g., YouTube.com/watch?v=rSnGJTxlo9M). Then read this poem aloud, emphasizing the sound words (*onomatopoeia*) especially.

2. **Invite students to chime in on the words that sound like sounds** while you read the rest of the poem aloud *(beep, roar, ruckus, bark, barking, clang, bang, Shhh, scritching, clinking, thumping).*

3. Stop everything and encourage students to rest their heads, close their eyes, and **tune in to the sounds all around them for one minute.** Then identify the sounds and talk about how each sound travels on waves to our ears.

4. Using the examples in this poem, **talk about how energy exists in many forms including sound waves.** Consider how the effects can vary with increasing or decreasing amounts of sound energy depending on how close the sound is and how much force is used to make the sound. Rank the examples in the poem from loud to soft: from the garbage truck *beep* to the *pop* of the lid coming off.

5. For another poem about sound waves, share **"Listen" by Amy Ludwig VanDerwater** (Kindergarten, Week 13, page 41). For more poetry about household sounds, look for *My House Is Singing* by Betsy Rosenthal.

## Week 14: Space

### Uh Oh, Pluto
by **Jeannine Atkins**

Once Pluto was proud to be called one of nine planets.
But astronomers decided he was too small,
too far from the Sun, made unpredictable orbits.
They tore pictures of poor Pluto off walls
and museum halls showed only eight planets.
Happily, Pluto found new friends, streaking balls
of rocks, dust, and ice called comets.
Orbiting whimsically together, Pluto is greatest of all!

### Take 5!

1. **Read this poem aloud and then share images and information about Pluto** and its place among the planets. Try SolarSystem.NASA.gov

2. Read the poem aloud again and **invite students to chime in on the planet name, *Pluto*,** each time it appears in the poem.

3. Talk about how **our understanding of science is changing all the time** as new information is gathered and new discoveries are made, particularly about space.

4. This poem provides **several key facts about our solar system; challenge students to list them** (once nine planets, now eight planets; Pluto's size, unpredictable orbits, distance from sun caused it to be reclassified as a dwarf planet) and then work together to identify the pattern of planets in the sky.

5. For a poem about another comet, seek out **"Comet Hunter" by Holly Thompson** (5th Grade, Week 14, page 242), and look for *Comets, Stars, the Moon, and Mars: Space Poems and Paintings* by Douglas Florian.

## WEEK 15: SUN, EARTH & MOON

### SECOND GRADE

### Take 5!

1. **Just for fun, stand tilted slightly while reading this aloud.** Afterward, you may need to explain key words in the poem, such as *axis, ecliptic, cryptic, astronomers*.

2. Read the poem aloud again, and **invite students to say the important last line together** (*The tilt causes seasons*) while you read the rest.

3. As students look at recognizable patterns in the natural world, **talk about the importance of weather and seasonal information for making choices in clothing, activities, and transportation**, particularly depending on where you are in the world at a given time.

4. **Show an explanatory video that demonstrates how earth's tilt** affects the seasons (e.g., YouTube.com/watch?v=Pgq0LThW7QA). Talk about how summer is in July in the northern hemisphere, but in December in the southern hemisphere.

5. Link this poem with another about the relationship between the earth and sun, **"The Shadow Grows (and Shrinks, and Grows)" by Laura Purdie Salas** (5th Grade, Week 15, page 243), and look for a poetry book by this same poet, *And Then There Were Eight: Poems about Space*.

## EARTH'S TILT
### by Douglas Florian

Earth's axis is tilted—
Its orbit ecliptic.
Exactly why
Remains somewhat cryptic.
Astronomers ponder
A whole host of reasons.
One thing is for sure—
The tilt causes seasons.

## WEEK 16: THE WATER CYCLE

### SECOND GRADE

### WATER ROUND
by **Leslie Bulion**

Raindrops fall
Splink
Splash
Onto soil
Drip
Seep
Into creeks
Trickle
Gurgle
Joining rivers
Whoosh
Flow
To the ocean
Go
Drops
Go!

Evaporate
Waft
Rise
Cool to droplets
Cloudy
Skies
Clouds grow heavier
And then?
Raindrops fall
Begin again!

### Take 5!

1. **Use an online audio sound library to play rain sounds in the background** while you read this poem aloud. One source is FreeSound.org.

2. Once again, **invite students to chime in on the words that sound like sounds** while you read the rest of the poem aloud *(splink, splash, drip, seep, trickle, gurgle, whoosh)*.

3. Work with students to **identify all the forms of water the poet includes:** *raindrops, creeks, rivers, ocean, skies.*

4. **Talk with students about how the poet captures the water cycle in just a few words.** Resources like the U.S. Geological Survey provide helpful models for exploring the processes in the water cycle, including evaporation, condensation, and precipitation, as connected to weather conditions. (See GA.water.USGS.gov/edu/watercycle.html.) And just for fun, look for the water cycle song at YouTube.com/watch?v=aU5WCqKHBfs.

5. Pair this with a similar poem by the same poet, **"Ocean Engine" by Leslie Bulion** (5th Grade, Week 16, page 244), and look for Marilyn Singer's book *How to Cross a Pond: Poems about Water.*

## Week 17: Weather & Climate

**Second Grade**

### Rain Gauge
by **Anastasia Suen**

I wonder how much it rained last night?
The rain gauge will show me the number.

I start at the bottom and count my way up.
I count up to the top of the water.

That's all? It rained for hours and hours.
I thought the number would be much higher.

Maybe tomorrow. I turn the gauge
upside down and pour out the water.

I wonder how much it will rain next time?
The rain gauge will show me the number.

### Take 5!

1. If possible, **show a rain gauge (or an image of a rain gauge) as your poetry** prop while reading this poem aloud.

2. Share the poem again, **but this time invite students to chime in on the question lines** (*I wonder how much it rained last night?; That's all?; I wonder how much it will rain next time?*) while you read the rest of the poem aloud.

3. For discussion: ***How does paying attention to weather reports help?***

4. This poem features a rain gauge, but other simple weather instruments include thermometers and wind vanes. **Talk about how these tools are helpful for measuring and recording weather information,** so we can make appropriate choices in clothing, activities, and transportation. One helpful resource is Weather.com.

5. Link this poem with another about measuring and reporting the weather, **"This Week's Weather" by Janet Wong** (3rd Grade, Week 17, page 165), and look for poems in *Weather: Poems for All Seasons* edited by Lee Bennett Hopkins. Connect these with the nonfiction picture book *Weather* by Seymour Simon.

**WEEK 18: FORCES OF NATURE**

SECOND GRADE

## PLATES
by **Ann Whitford Paul**

We live on plates—
not the eating-off-of kind—
great slabs of rock
that slowly move
deep
    deep
        deep beneath our feet.

Sometimes some plates—
not the eating-off-of kind—
collide . . . slide . . .
trembling the very ground we walk—
EARTHQUAKE!

Everything shakes.
Bridges split,
chimneys crack,
towers crumble,
and inside houses,
plates—
the eating-off-of kind—
shatter,
break.

**Note:** Tectonic plates move in an area called the *lithosphere*, which includes the earth's upper mantle and crust. This movement is due to pressure and heat from the mantle (convection currents).

### Take 5!

1. **The logical prop for sharing this poem is a simple dinner plate or paper plate.** Have one ready to show as you read this poem aloud, pausing dramatically at the end of each line. Want to make a pop-up book demonstrating plate tectonics? Find instructions at EarthPopUpBook.weebly.com.

2. **For a follow-up reading, invite students to read the lines** *not the eating-off-of kind* and *the eating-off-of kind* while you read the rest of the poem aloud.

3. **Time to discuss emergency preparedness,** whether you live in an earthquake-prone area or not. Review the school's plan together.

4. **Talk with students about the earth's different layers** (crust, mantle, outer core, and inner core) and briefly introduce the concept of tectonic plates (*great slabs of rock / that slowly move . . .* and sometimes *collide . . . slide . . .* ). Provide visuals with this excellent video from National Geographic, "Earthquakes 101" at YouTube.com/watch?v=VSgB1IWr6O4.

5. For a completely different look at earthquakes, share **"Future Dreams Idea #63" by Janet Wong** (Kindergarten, Week 36, page 64). Also look for *Earthshake: Poems from the Ground Up* by Lisa Westberg Peters, and Seymour Simon's nonfiction book *Earthquakes*.

## Week 19: Soil & Land

**Second Grade**

### Take 5!

1. Before sharing this poem, take a moment to **encourage students to close their eyes and imagine a serpent** and picture a large creature with a long curving tail. You may need to define *serpent*—snake, dragon, monster, etc. Then continue by reading this poem aloud.

2. **Work together to research images of and facts** about glaciers. One excellent source is the National Snow and Ice Data Center at NSIDC.org/cryosphere/glaciers.

3. Challenge students to work in pairs or trios to **draw a picture of how they picture the serpent in the poem,** then post the pictures alongside images of glaciers and a copy of the poem.

4. Use this poem to talk with students about **how the Earth consists of natural resources and how its surface is constantly changing.** The poet compares a glacier to a serpent and refers to the eroded landform (*valley trail*) left behind by glaciers.

5. Pair this poem with another about changes in the earth's surface, **"Riddle for a Wet Day" by Irene Latham** (1st Grade, Week 18, page 86). Also look for selections in *Once Upon Ice and Other Frozen Poems* by Jane Yolen, and the nonfiction picture book *Glaciers* by Colleen Sexton.

## Glacier
### by Kate Coombs

The great ice serpent
dragged its cold coils,
scraping its tail
against rocks,
leaving its sign behind,
a curving valley trail.

## Week 20: Natural Resources

### Fossil Fuels
by **Janet Wong**

They're talking
about fossil fuels
on the news.
I ask Pop
what those are
and he says
fossil fuels
are oil and gas and coal
made from plants and plankton
that sank down in the water
and got covered and cooked
in a thick mud crust
for millions and millions
of years.

No wonder
it costs so much
to fill up our car:
our fuel
took millions and millions
of years
to make!

---

### Take 5!

1. **Read this descriptive poem aloud, and then show a diagram of how fossil fuels are formed.** One great source is the U.S. Energy Information Administration at EIA.gov/kids/energy.cfm?page=oil_home-basics.

2. Share the poem again, and this time **invite students to be the child in the poem and say those lines**—*I ask Pop / what those are / and he says*—while you read the rest of the poem aloud.

3. **Talk with students about alternative modes of transportation** (cars, buses, subway, bicycles, Segways, walking, etc.).

4. Use this poem to talk about how the natural world provides fuel for our manmade machines. **Discuss the different possible ways to fuel a car and whether they rely on fossil fuels or not** (e.g., gasoline, diesel fuel, electricity, a hybrid combination, natural gas, ethanol, etc.). One helpful site is EPA.gov/otaq/fuels/alternative-renewablefuels/.

5. Pair this poem with another by the same poet, **"Auntie V's Hybrid Car" by Janet Wong** (Kindergarten, Week 20, page 48), and with *Behind the Wheel: Poems about Driving* also by Wong.

## WEEK 21: ECOSYSTEMS

**SECOND GRADE**

### THE RAIN FOREST
by **Bobbi Katz**

It's a weaving—growing, breathing.
Huge trees form a canopy.
Within this leafy rooftop,
there's more life than we can see.
Above, a few trees poke through,
where Harpy Eagles look for prey.
The rain forest restaurant
serves them banquets night and day.

In the pulsing understory
high above the forest floor,
there are striders, swingers, gliders—
jaguars, monkeys, bats, and more.
There are countless birds and insects,
we cannot name them all.
Fruits and flowers, butterflies—
flying,
       crawling,
chewing,
       calling—
clicks and whispers—
screeching cries.

### Take 5!

1. Set the stage for this vivid poem by **showing a short videoclip of a rainforest scene complete with animal sounds.** One example is at YouTube.com/watch?v=JZc94cO54kY. Then read the poem aloud slowly and dramatically.

2. For a follow-up reading, **invite student to choose their favorite animal mentioned in the poem and to say the animal name** when it appears in the poem as you read the rest aloud. Animals include *Harpy Eagles, jaguars, monkeys, bats, birds, insects, and butterflies.*

3. **Talk with students about organizations that help protect rainforests** and other ecosystems, like the World Wildlife Federation (Panda.org).

4. **Use this poem to look at how various living organisms depend on each other and on their environments** for survival, and consider what happens when those ecosystems are destroyed or disrupted. Work together to research the layers of the rainforest and which creatures inhabit each.

5. For another poem on this important topic, look for **"Tropical Rain Forest Sky Ponds" by Margarita Engle** (4th Grade, Week 21, page 209), and share selections from *Chatter, Sing, Roar, Buzz: Poems about the Rain Forest* by Laura Purdie Salas.

## WEEK 22: ADAPTATIONS & TRAITS

SECOND GRADE

### THE LEOPARD CANNOT CHANGE HIS SPOTS
by **Lesléa Newman**

The leopard cannot change his spots
Into stars or polka-dots.
The tiger cannot change her stripes
("I wish I could," she sometimes gripes.)
The scales remain upon the fish
Though that is sometimes not his wish.
The kangaroo can try and try
But she will never learn to fly.
The cat can't bark, the dog can't purr,
The rabbit cannot change her fur.
The frog can leap but he can't walk,
The lark can sing but she can't talk.
Since we can't be what we are not,
Let's all be grateful for our lot.

### Take 5!

1. Before sharing this poem, challenge students to **imagine what each animal might look like** as they listen to you read the poem aloud.

2. Once again, in a follow-up reading, **invite student to choose their favorite animal mentioned in the poem and to say the animal name** when it appears in the poem as you read the rest of it aloud. Animals include *leopard, tiger, fish, kangaroo, cat, dog, rabbit, frog, lark.*

3. Challenge the students to work in pairs or trios to **draw a picture of one of the make-believe versions of an animal in the poem,** and then post all the pictures in order corresponding to the poem.

4. **Use this poem to contrast the normal with the imagined qualities for each animal,** and talk about how the physical characteristics and behaviors of animals help them meet their basic needs (e.g., fins help fish move and balance in the water).

5. Connect with another poem about animal attributes, **"Snake Traits" by Linda Ashman** (Kindergarten, Week 22, page 50). Or just for fun, share Jack Prelutsky's poems about imaginary "hybrid" animals in the books *Scranimals, Behold the Bold Umbrellaphant,* and *Stardines Swim High Across the Sky.*

## Week 23: Cycles

**Second Grade**

### Becoming Butterflies
by **Jeannine Atkins**

One tiny egg breaks.
A caterpillar comes out,
eats her former shell,
and sheds her skin.

The caterpillar eats more,
grows more. Her skin splits again,
before she spins
a safe, still chrysalis.

At last, a butterfly breaks through.
She unfolds wet wings,
flutters, seeks nectar,
flits, catches another's notice.

Finally (or is it first?), she finds
a leaf where she lays new eggs,
small as pencil points.
. . . One tiny egg breaks.

### Take 5!

1. If possible, **make a tiny mock butterfly** by cutting tissue paper into a butterfly shape and hold it gently in your hand as you read this poem aloud.

2. **Divide the class into four groups** and invite each group to join you in reading ONE of the stanzas aloud.

3. Use this poem to consider some of the unique stages that insects undergo during their life cycle. **Challenge the students to work in pairs or trios to draw a picture for one stanza,** and then post all the pictures in order corresponding to the poem.

4. **Talk with students about each stage of the life cycle of a butterfly outlined in the poem** (egg, caterpillar/chrysalis, butterfly, new egg). Share a time lapse video that captures the process at YouTube.com/watch?v=7AUeM8Mbalk.

5. For another poem about an insect, share **"Cicada / Chicharra" by Guadalupe Garcia McCall** (5th Grade, Week 23, page 251). Also look for Joyce Sidman's *Butterfly Eyes and Other Secrets of the Meadow* and Avis Harley's *The Monarch's Progress: Poems with Wings.*

## WEEK 24: PATTERNS

### RINGS NOT LETTERS
by **Juanita Havill**

A tree writes the story of its life
in rings not letters.
One tiny ring at the center:
"Here is where I began."
Next year a new ring:
"Look how much I grew."
Wide bands between rings:
"Hooray for rain and sun."
Narrow bands:
"It's hot and dry and I'm so thirsty."
Fires, insects, the weight
of a fallen tree against the trunk,
all written in rings, not letters,
the life story of a tree.

**Note:** This poem was inspired by information from the Tree Ring Lab at the University of Arizona (LTRR.Arizona.edu) and an image and explanation at Arborday.org/trees/ringsLivingForest.cfm.

### Take 5!

1. Read this poem aloud, pausing for each response in quotation marks. **Then show a cross-section image of a tree revealing its rings.** One interactive example is at ArborDay.org/kids/carly/lifeofatree/.

2. Read the poem together, **inviting students to read the portions within quotation marks**—*Here is where I began; Look how much I grew; Hooray for rain and sun;* and *It's hot and dry and I'm so thirsty*—while you read the rest aloud.

3. Work together to draw a diagram of the rings of a tree and to **label the rings using the words and lines from the poem.**

4. **Use the details in this poem to point out the recognizable patterns in the natural world.** In particular, talk about how the tree's growth reflects the weather it has experienced, including rain, sun, heat, drought, fire. Identify what all plants and trees need (sun, rain, soil).

5. For more poems about trees, look for **"Windfall in The Andrews Forest" by Joseph Bruchac** (4th Grade, Week 23, page 211) and selections from *Poetrees* by Douglas Florian, *Winter Trees* by Carole Gerber, and *Forest Has a Song* by Amy Ludwig VanDerwater.

## Week 25: Human Body

SECOND GRADE

### Take 5!

1. Read this poem aloud and then take a moment to define any unfamiliar words like *vaccination* or *immunity*.

2. Share the poem again, and this time **invite students to say the "reaction" lines**—*please make it quick* and *think I'll be sick*—while you read the rest of the poem aloud.

3. Point out that sometimes our bodies need help staying healthy and **talk about what kinds of things we can do to stay well** (hand washing, balanced meals, regular exercise, plenty of sleep, etc.).

4. This poem personalizes the need for vaccinations and immunizations, but many scientists enter this field to study diseases and how to prevent them. **Talk with students about what a scientist is and explore what different scientists do** in the field of health and medicine and beyond. Introduce names like Edward Jenner, Alexander Fleming, and Jonas Salk.

5. For another poem about personal health, look for **"Seeing School" by Kate Coombs** (1st Grade, Week 25, page 93).

### SHOTS! SHOTS! SHOTS!
by **Joy Acey**

I need the shots,
please make it quick.
I see a needle,
think I'll be sick.
You say it is just
a little stick.
I know vaccinations
build immunity
but getting them
takes bravery.

## Week 26: Kitchen Science

### Mold
#### by Charles Waters

When food becomes too old
It's visited by mold,
Fuzzy green, blue, black gunk.
Please, throw it out; it's junk.
Although some mold is helpful, yes—
It's hard for you or me to guess
So, please be safe: toss it! Why?
You could get sick from fungi.

### Take 5!

1. **Read this poem slowly, pausing before the final word.** Then talk about the word *fungi* and define it for students (an infectious organism).

2. Read the poem aloud again, and **invite students to say the important last word together** (*fungi*) while you read the rest of the poem.

3. For discussion: **What kinds of science experiments can we do in the kitchen** (with mixing and cooking, growing plants, studying germs, etc.)?

4. Use this poem to **talk about how living organisms depend on each other and on their environments**. How does mold grow on food and how do we avoid it? (Keep food fresh and refrigerated or packaged properly and dispose of it when it expires.) It can be interesting to moisten a slice of white bread and seal it in a plastic bag. Watch the fungi grow (then dispose of it without opening the seal).

5. For a more positive poem about the potential benefits of food rotting, look for **"Pumpkin Experiment" by Mary Lee Hahn** (1st Grade, Week 6, page 74).

## Week 27: Video Technology

**SECOND GRADE**

### Take 5!

1. **Display the homepage or screenshot of a popular video game or app in the background** as you read the poem aloud.

2. The speaker in the poem lists all the tasks that make the aunt's job as a game programmer special. **Invite students to chime in on the fun, final line of the poem** *(it's never, ever boring!)* while you read the rest of the poem aloud.

3. **Talk with students about their favorite video games** and what features they particularly like about them.

4. **Highlight the computer science terminology used in this poem** and talk about what these words mean (*game programmer, programs, video, commands, computer*).

5. Look for a parallel poem about making video games with **"Computer Geek / Compu-nerdo" by Carmen T. Bernier-Grand** (Kindergarten, Week 33, page 61).

## Game Programmer
by **Janet Wong**

My aunt has the best job ever.
She programs video games.
Someone else
comes up with the stories.
Someone else
comes up with the names.
But she puts all the commands in.
She makes the games work right—
so cars will move
when you want them to,
so soldiers can see at night.
She speaks to the computer
and calculates the scoring.
My aunt has the most incredible job—
it's never, ever boring!

## WEEK 28: MACHINES

### GEARS
by **Michael Salinger**

A gear is a machine
that needs only two parts.
Like wheels with teeth,
when one spins the other starts
to turn in what is called a ratio.
Gears come in all
different shapes and sizes,
mostly doing their work
inside of stuff.
Where might we use
some gears today?
What spins or turns?
What rotates or grinds?
What lifts or what lowers?
How many gears can you find?

### Take 5!

1. **Read this poem aloud while showing a video of gears at work, without sound.** There is a helpful example at YouTube.com/watch?v=WYcqJ5HdxA4.

2. Share the poem again, **but invite students to choose a question line to chime in on** (any of the last five lines) while you read the rest of the poem aloud. Different students can choose different lines.

3. Tackle one of the questions in the poem together, such as *How many gears can you find?*

4. **Use this poem to talk about the attributes of gears** (*two parts, wheels with teeth, spins, different shapes and sizes*) and the patterns of movement identified here (spinning). Work together to research possible everyday uses of gears.

5. Look for another mechanically-minded poem by this same poet, **"Levers" by Michael Salinger** (1st Grade, Week 28, page 96), as well as **"The Crane Operator" by Rebecca Kai Dotlich** (3rd Grade, Week 29, page 177).

## WEEK 29: BUILDING THINGS

### SECOND GRADE

### MY ROBOT
by **David L. Harrison**

I built my first robot with sticks and mud.
I gave it a pebble brain.
Turned out to be a dud.
You might say that experiment was in vain.
It dissolved when I left it in the rain.

The second one was made of papier-mâché.
It wasn't much of a robot,
Though I thought of it that way.
Mostly it just sat there on my table
Being as much robot as it was able.

The third robot I built was from a kit.
It had a metal frame
And a motor came with it.
It looked really truly like a robot.
Even so, it couldn't do a lot.

One of these days I'm going to get it right.
The robot in my future
Will be a glorious sight.
My robot will follow me around.
My robot! What an awesome sound!

### Take 5!

1. **Read this poem aloud, pausing briefly between stanzas.** Then follow up with some exploration of this excellent robotics website at Robotics.usc.edu/~agents/k-12/index.php.

2. Share this poem again, and **invite students to join in on the first line of each stanza** as you read the rest aloud.

3. Collaborate with students to **create a quick glog, a digital interactive poster (using Glogster.com), pulling together images for each stanza of the poem**. Show the students the choices of text, fonts, color, graphics, and even animation, if possible, while you input those items and create the finished product.

4. **Use this poem to focus on the model of scientific problem solving it presents.** Talk about each problem and how it was solved or improved upon. Although these are silly examples, scientists use the same critical thinking and analysis to identify and explain a problem and propose a solution. Challenge students to build on the poem by brainstorming future robot possibilities.

5. For another poem about trial and error in science, revisit **"My Experiment" by Julie Larios** (2nd Grade, Week 6, page 114).

## Week 30: Science Fair

### Second Grade

### Science Project
by **Lee Wardlaw**

O, Dormant Cone, fear
not! I am Pélé, Goddess
of Fire, who rules your

throat and ire. Awake!
Swallow my potion! Release
what smolders, sickens,

below. You choke. Cough—
burble scarlet froth! My class
erupts in applause.

### Take 5!

1. **Set the stage for this poem by showing an image or a video** of a volcano. Then read the poem aloud with an overly dramatic voice. Share "Volcanoes 101" from Video.NationalGeographic.com. Talk with students about some of the challenging words: *dormant, Pélé, ire, smolders, burble, scarlet, froth, erupts.*

2. **Invite students to provide sound effects for** the final stanza of the poem (choking, coughing, applauding) as you read the poem aloud again, dramatically slowing down for the final stanza.

3. **Here the poet uses the haiku poem form** to describe a mock volcano science project. Talk about different ways one could describe a science project (haiku poem, list form, note cards, etc.).

4. Talk with students about how **the project in this poem provides an example of *chemical reactions.*** (You can't separate the homemade lava back into the original baking soda and vinegar.) Work together to research how mock "lava" can be created for a pretend volcano. Here's one example video: YouTube.com/watch?v=GNaJDSk2eoE.

5. For another poem about mixing things, look for **"Breakfast Alchemy" by Mary Quattlebaum** (3rd Grade, Week 26, page 174), or look for selections from *Volcano Wakes Up!* by Lisa Westberg Peters.

## WEEK 31: FAMOUS SCIENTISTS

**SECOND GRADE**

### Take 5!

1. Read this poem aloud with feeling, and then **show the short videoclip of Jane Goodall herself making the chimp call at** Youtube.com/watch?v=2jF0Hs7tIMA.

2. Share the poem again, and **invite students to applaud whenever the word *clap* appears** in the poem while you read it aloud.

3. Scientists do all kinds of work, including studying and living with animals. **Talk about some of the pros and cons of studying animals firsthand.**

4. **Work together to explore Goodall's work** through the Jane Goodall Institute at JaneGoodall.org. What has she accomplished and what does she still want to do? Send an e-card with your favorite chimpanzee photo (look under the "Support" link). Take a class poll to determine which chimpanzee to choose, graph the results, and send the e-card with the winning image.

5. Follow up with a visit to the natural habitat of chimpanzees, the rain forest, by rereading **"The Rain Forest" by Bobbi Katz** (2nd Grade, Week 21, page 129), and then look for the picture book biography *Me...Jane* by Patrick McDonnell.

### JANE GOODALL BEGINS A SPEECH
by **Susan Marie Swanson**

Picture a grand auditorium
lit with glittering chandeliers.
A scientist steps up to the microphone.
"Good morning," she says. "Listen."
The place is hushed.

Then, she starts to hoot and whoop,
loud and louder, shrieking,
turning the whole room wild!
From a forest in Tanzania
she has brought a greeting
from the wild chimpanzees!

The crowd claps and claps
for Jane Goodall
and her chimpanzee hello.
They clap for the stories she will tell
about the chimps at Gombe
and how she became a field biologist
who carries chimpanzee voices
all over the world.

## Week 32: More Famous Scientists

**Second Grade**

### Wondering Why
*Charles Darwin (1809-1882)*
  by **Shirley Smith Duke**


Darwin sailed as a naturalist on a far sea journey—
the *Beagle* was his ship.
He gathered samples of the different life he saw
from the Galapagos Islands off South America.

He took careful notes and then drew.

Different beaks on finches for each island.
Different shells on tortoises on the islands.
He thought and thought about why they looked alike,
yet had some noticeable differences.

Darwin had many questions.

Why do differing groups live in different places?
Do the strongest and best fit survive?
Can the young get their traits from their parents?
What makes species change over time?

Natural selection?

Darwin gathered up tortoises, fifty of them,
to take back to England to study.
But, as the ship sailed home, sailors ate turtle soup—
thank goodness for Darwin's drawings and notes!


**Note:** Darwin's theory of evolution by natural selection, controversial when introduced because the idea appeared to contradict the prevailing religious idea on how the world was created, has since become accepted by science.

### Take 5!

1. Read the poem aloud with a brief pause at the end of each stanza. Then take a moment to **locate the Galapagos Islands on a world map** using Google Maps, complete with video footage.

2. For a follow-up reading, **invite students to read the single line stanzas** (*He took careful notes and then drew; Darwin had many questions; Natural selection?*) while you read the rest of the poem aloud.

3. **Work together to learn three facts about Charles Darwin**. Check out the resources and online exhibits of the American Museum of Natural History at AMNH.org/exhibitions/past-exhibitions/Darwin.

4. **Talk with students about Darwin's model of studying nature with a notebook and drawings.** If possible, guide students in brief outdoor observations with notebooks in hand. Encourage them to ask and write down questions about organisms, objects, and events during observations and investigations. You may want to narrow the choice to a specific object outside to observe, like a tree or a patch of flowers.

5. For another poem about why we are the way we are, share **"Inherit Tense" by Charles Ghigna** (Kindergarten, Week 24, page 52). Also look for the nonfiction books *One Gorilla: A Counting Book* by Anthony Browne and *Life on Earth: The Story of Evolution* by Steve Jenkins.

## WEEK 33: COMPUTERS

**SECOND GRADE**

### (SUPER)POWER: (TO THE)POINT
by **Kristy Dempsey**

Flash! Shazam! I slide onscreen,
designed to grab attention.
Nothing more and nothing less
than what deserves a mention.
Clear and focused, on the scene,
I'm breaking down this joint,
brandishing my superPower:
Point by Point by Point.

### Take 5!

1. Read the poem aloud with a brief pause at the end of each line. Then take a moment to **show how PowerPoint (or similar) slideshow software works**.

2. Share the poem again, and **invite students to start things off by shouting the words *Flash! Shazam!*** in the poem while you read the rest of the poem aloud.

3. **Work together to create a simple PowerPoint slideshow.** Consider using this poem to create a slideshow that features an image paired with each line from the poem in eight separate slides. Then read the poem aloud together while showing the slides.

4. **Talk with students about how scientists use all kinds of tools for their investigations and for sharing their findings.** Some tools like computers help in the research and investigation process, and some (like PowerPoint and other similar computer programs) can assist in showing their results.

5. For more poems about using computer software, look for **"My Photo Experiment" by Janet Wong** (4th Grade, Week 33, page 221) or **Computer Geek / Compu-nerdo by Carmen T. Bernier-Grand** (Kindergarten, Week 33, page 61).

## WEEK 34: SCIENCE CAREERS

**SECOND GRADE**

### WATER ENGINEERED
by **Sara Holbrook**

Water is pumped all around.

Engineers plan
how it's captured,
channeled,
and hosed
through huge valves,
   (open and closed)
  designed to make water rush
when I need a shower,
  a drink,
    or a flush.

Water is pumped all around.

### Take 5!

1. Read this poem aloud slowly, emphasizing each line **while you play this 3D video of water resource recovery (without sound) as your backdrop:** Youtube.com/watch?v=A2FmNrEmowE.

2. For a follow-up reading, **invite students to join you in reading the repeated first and last lines** while you read the rest of the poem solo.

3. Look for David Macaulay's *How It Works* books like *Toilet: How It Works* for talking points, or **explore the excellent resources at GetCaughtEngineering.com.**

4. Engineers use science for problem solving and making decisions. **Talk about the many possible science careers** listed at ScienceBuddies.org.

5. For more water-themed poems, look for **"Water" by Kate Coombs** (Kindergarten, Week 35, page 63) and selections from *Splash! Poems of Our Watery World* by Constance Levy.

## Week 35: Future Challenges

**SECOND GRADE**

### The Lament of Lonesome George
by **Jane Yolen**

*"Lonesome George is gone, and there will never be another like him."*
—*New York Times* obituary, July 2, 2012

My name is George and I have lived
a long and simple life.
The last of all my tortoise kin,
no children and no wife.

My quiet archipelago
holds all my simple needs.
Just living to a hundred years
is reckoned in my deeds.

My keepers are my only friends,
as I close down my race.
"A Giant Tortoise" papers said
"extinction has a face."

But as I face extinction,
this one small truth I see:
You humans are a lot of you,
but I was one of me.

### Take 5!

1. Read the poem aloud while **playing a video of Lonesome George in the background.** Use YouTube.com/watch?v=ykIK05ysDuc.

2. Share this poem again, and **invite students to read the powerful last line** while you read the rest of the poem aloud.

3. **Talk about key words in the poem,** like *tortoise, archipelago, reckoned, extinction.*

4. We know that all living organisms have basic needs that must be met for them to survive within their environment. **Work with students to research which animals are on the endangered species list. Note that scientists are exploring the possibility that Lonesome George might not be the last of his kind after all.** This site has an interactive map showing endangered animals by state: FWS.gov/endangered/map/index.html.

5. Read another poem about an endangered animal, **"Rocky Rescue" by Robyn Hood Black** (4th Grade, Week 34, page 222), or selections from Jane Yolen's poetry book about ancient animal breeds, *The Originals: Animals that Time Forgot;* Tony Johnston's *An Old Shell: Poems of the Galapagos;* and Janet Wong's *Once Upon a Tiger: New Beginnings for Endangered Animals.*

## WEEK 36: FUTURE DREAMS

**SECOND GRADE**

### SPACE YACHT
by **Juanita Havill**

Imagine a space ship with sails,
a solar yacht that travels far
and carries you from Earth to Mars
on streams of photons from our star.

Never-ending streams of photons,
particles in waves of light,
bounce off sails to push your space ship
like blowing wind upon a kite.

Wind-powered kites attached to string
guided by humans on the ground
can never fly above the clouds.
Not so your yacht—it's universe bound.

**Note:** From Space.com: "The largest solar sail ever constructed is headed for the launch pad in 2014 on a mission to demonstrate the value of 'propellantless propulsion'—the act of using photons from the sun to push a craft through space."

### Take 5!

1. Before sharing this poem, take a moment to **encourage students to close their eyes and imagine a space ship that could take us all to Mars**. Then continue by reading this poem aloud. You may need to explain a few key words like *yacht, photon, particles*.

2. For a follow-up reading, **invite students to chime in on those three key words,** *yacht, photon,* and *particles*, while you read the rest of the poem aloud.

3. Work together to **create a drawing of what the "Space Yacht" in this poem might look like**. Talk about what students would want to have on board to take along to Mars.

4. Students know that energy exists in many forms. This poem mentions solar and wind energy with words like *photons, / particles in waves of light* and *Wind-powered kites*, for example. **Use this poem to talk about these two forms of energy: solar and wind.** Look for information from the U.S. Energy Information Adminstration at EIA.gov/kids/energy.cfm?page=2.

5. For another poem with a kite connection, revisit **"Go Fly a Kite" by Laura Purdie Salas** (2nd Grade, Week 12, page 120), or look for another future forecasting poem, **"Everyday Astronaut" by Carmen Tafolla** (1st Grade, Week 36, page 104).

*The Poetry Friday Anthology for Science*

# Poems for Third Grade

# NGSS Science and Engineering Practices: Third Grade

*These practices form the foundation of disciplinary literacy in science and integrate reading, writing, listening, and speaking skills from the language arts. Here we indicate which weekly poems emphasize which science and engineering practices at each grade level.*

| PRACTICE | POEM |
| --- | --- |
| Asking questions and defining problems | Weeks 1, 2, 3, 34, 35 |
| Developing and using models | Weeks 15, 20, 29 |
| Planning and carrying out investigations | Weeks 5, 6, 30 |
| Analyzing and interpreting data | Weeks 8, 9, 13, 19, 21, 26, 32, 33 |
| Using mathematics and computational thinking | Weeks 7, 17 |
| Constructing explanations and designing solutions | Weeks 11, 12, 16, 25, 27, 28, 36 |
| Engaging in argument from evidence | Weeks 10, 22, 31 |
| Obtaining, evaluating, and communicating information | Weeks 4, 14, 18, 23, 24 |

## Third Grade

| | | |
|---|---|---|
| week 1 | Scientific Practices | Which Ones Will Float? *by Eric Ode* |
| week 2 | Lab Safety | Things to Do in Science Class *by Laura Purdie Salas* |
| week 3 | Ask and Ask Again | Inquiry *by Cynthia Cotten* |
| week 4 | Observations | Classroom in the Meadow *by Jeannine Atkins* |
| week 5 | Predictions & Hypotheses | Testing My Hypothesis *by Leslie Bulion* |
| week 6 | Investigations | Meet Mr. Wizard *by George Ella Lyon* |
| week 7 | Data | Zapped! *by April Halprin Wayland* |
| week 8 | Tools of Science | Armor *by Margarita Engle* |
| week 9 | Matter | Questions That Matter *by Heidi Bee Roemer* |
| week 10 | More Matter | The Brink *by Janet Wong* |
| week 11 | Force, Motion & Energy | After I Made a Huge Mess… *by Mary Lee Hahn* |
| week 12 | More FM&E | Lift *by Marilyn Singer* |
| week 13 | Light & Sound | Sound Waves *by Michael Salinger* |
| week 14 | Space | Orion Nebula *by Mary Lee Hahn* |
| week 15 | Sun, Earth & Moon | Lunar Eclipse *by Bobbi Katz* |
| week 16 | The Water Cycle | We Need Green Seaweed! *by Margarita Engle* |
| week 17 | Weather & Climate | This Week's Weather *by Janet Wong* |
| week 18 | Forces of Nature | Tornado! *by Carole Gerber* |
| week 19 | Soil & Land | Trilobite *by Mary Ann Hoberman* |
| week 20 | Natural Resources | What Makes a Turbine Turn *by Steven Withrow* |
| week 21 | Ecosystems | Tide Pool *by Jane Yolen* |
| week 22 | Adaptations & Traits | Camouflage *by Margarita Engle* |
| week 23 | Cycles | Sun-Kissed/Besado por el sol *by Guadalupe Garcia McCall* |
| week 24 | Patterns | Citizen Scientist *by Shirley Smith Duke* |
| week 25 | Human Body | Protecting My Friend *by Jacqueline Jules* |
| week 26 | Kitchen Science | Breakfast Alchemy *by Mary Quattlebaum* |
| week 27 | Video Technology | What Am I? *by Esther Hershenhorn* |
| week 28 | Machines | Five O'Clock Rush/Prisas a las cinco *by F. Isabel Campoy* |
| week 29 | Building Things | The Crane Operator *by Rebecca Kai Dotlich* |
| week 30 | Science Fair | Science Fair Project *by Eric Ode* |
| week 31 | Famous Scientists | Considering Copernicus *by Bobbi Katz* |
| week 32 | More Famous Scientists | Galileo Galilei *by Renée M. LaTulippe* |
| week 33 | Computers | Wiki Alert *by Debbie Levy* |
| week 34 | Science Careers | Invention Intentions *by Kristy Dempsey* |
| week 35 | Future Challenges | Cancer *by Mary Lee Hahn* |
| week 36 | Future Dreams | Moving to Atlantis City, 2112 *by Steven Withrow* |

The Poetry Friday Anthology for Science

"Logic will get you from A to B.
***Imagination will take you everywhere.***"

≈℘ Albert Einstein ℘≈

## Week 1: Scientific Practices

### Take 5!

1. **If you have a cell phone with a "Siri" kind of function, try asking it one of the questions in this poem** just for fun before reading this poem aloud. Or simply draw a giant question mark on the board before launching into reading the poem.

2. Share the poem again, and **invite students to chime in on the first two and last two lines of each stanza (which are repeated)** while you read the rest of the poem aloud.

3. **If possible, try this poem experiment together** by filling a small container with water and seeing if a few small everyday objects float or sink in it. Guess the outcome before trying each object.

4. Use this poem to talk with students about how scientists engage in questioning, critical thinking, and scientific problem solving. **Work together to make a list of the properties that are examined in this poem** (floating, sinking, size, shrinking, color, odor). Talk about how it's important to verify data through multiple trials and consistent results.

5. Pair this poem with another testing the properties of objects, **"Sink or Float" by Janet Wong** (Kindergarten, Week 5, page 33), and look for more science-themed poems in *Spectacular Science: A Book of Poems* edited by Lee Bennett Hopkins.

## Which Ones Will Float?
### by Eric Ode

Which ones will float?
Which ones will sink?
Which will grow larger?
Which ones will shrink?
We'll test and investigate,
watch and compare.
But will we agree
with the answers we share?
Maybe we won't,
and then, if we don't,
we'll try, try again
and decide what to think.
Which ones will float?
Which ones will sink?

Which will turn purple?
Which will turn pink?
Which will smell pleasant?
Which ones will stink?
We'll study the data
and enter our claim.
Maybe our answers
will look just the same.
But maybe they won't,
and then, if they don't,
we'll try, try again
and decide what to think.
Which will turn purple?
Which will turn pink?

## WEEK 2: LAB SAFETY

**THIRD GRADE**

### THINGS TO DO IN SCIENCE CLASS
by **Laura Purdie Salas**

**L**ook at labels.
**A**sk advice.
**B**e sure to check directions twice! Wear

**S**olid shoes to shield your feet,
**A**nd keep your table clean and neat.
**F**ollow rules that you are given.
**E**xplore
**T**he startling world
**Y**ou live in.

### Take 5!

1. **Read this poem aloud while pantomiming some of the actions in the poem** (looking at a label, showing your shoes, straightening items on the table).

2. Use the acrostic formula for this poem to **invite students to say the first letter of each line** while you read the rest of the poem aloud.

3. Take each letter in the word *SCIENCE* and **challenge students to collaborate on creating another acrostic poem that highlights science**.

4. **Here is the teachable moment for talking about the importance of safety in science.** Break down the poem and list all the safety issues included. Research your local safety standards and discuss with students the value of following directions, keeping the area clean, wearing safety goggles, washing hands, and using materials appropriately.

5. Connect this poem with another that highlights science lab safety, **"Dinos in the Laboratory" by Kristy Dempsey** (4th Grade, Week 2, page 190).

## Week 3: Ask and Ask Again

### Third Grade

### Take 5!

1. Set the stage for this poem by **drawing a giant question mark on the board.** Then read the poem aloud, pausing at the end of each question line.

2. Alert students to all the question lines in this poem (*Who?, What?, Where?, When?*, etc.). Display the text of the poem, and **invite students to choose their favorite question line and chime in** when that line appears.

3. **Start a question wall** (with craft paper on a door or bulletin board) and invite students to brainstorm and add their own questions to ponder, beginning with some of these.

4. Use this poem to focus on how scientists use questioning and inquiry to launch scientific investigations. **Make a list of the questions presented here and talk about how they help us begin the research process** through observation, data collection, and more question asking. Questions include *Who?, What?, Where?, When?, Why?, What if?, Could I try?, Can I test?, Can I measure?, Can I describe?*

5. Follow up with another poem filled with questions, "**Late Night Science Questions**" **by Greg Pincus** (2nd Grade, Week 3, page 111), or "**Paper Airplanes**" **by Janet Wong** (5th Grade, Week 3, page 231). Or look for selections from *Where Fish Go in Winter: And Answers to Other Great Mysteries* by Amy Goldman Koss.

## Inquiry
by **Cynthia Cotten**

There are so many questions
I could ask.
Which one
is the right one?

Some questions are simple—
Who?
What?
Where?
When?

But simple questions usually have
simple answers.
They take me only so far.

There are others, though—
Why?
What if . . . ?
Could I try . . . ?
Can I test?
Can I measure?
Can I describe?

These questions
lead to observations—
to answers that go deeper,
further,
that might even lead to
more questions.

I've heard that no question
is wrong
if you don't know the answer.
I guess the right one
depends on what you want
to know.

## WEEK 4: OBSERVATIONS

### THIRD GRADE

## CLASSROOM IN THE MEADOW
### by Jeannine Atkins

Children crawl through thickets
to help their father, Professor Fabre.
Their quick hands catch crickets,
carefully tuck bugs in old snail shells.

The children crouch over mules' footprints, small
rain-filled pools, and watch mayflies swarm.
"Observe," Papa says. "Patience shows all!"
They learn that butterflies taste leaves with their feet.

They measure the leaping of fleas
and the vaulting of grasshoppers.
Listen! Crickets hear with their knees
and chirp with their wings.

The children run through fields with cheers for creatures
who are hunters, builders, weavers, miners, architects,
and engineers. Insects are also teachers,
giving lessons in palms, puddles, and mid-air.

**Note:** Jean-Henri Fabre (1823-1915) is considered a founder of modern entomology, the study of insects. When his seven children grew up, one son continued to work with him, while those who left for other villages or cities in France sent home insect specimens.

### Take 5!

1. Before reading this poem aloud, challenge students to **listen for all the different insects** included in the poem (*crickets, bugs, mayflies, butterflies, fleas, grasshoppers*).

2. As you share the poem again, focus on the many roles insects play in the natural world. You read the poem aloud and **invite students to chime in on these words in the final stanza:** hunters, builders, weavers, miners, architects, engineers, teachers.

3. **Work together to research details about the actual person featured in the poem**, Jean-Henri Fabre (pronounced **Fob**-ruh). Use this source dedicated to his life and work: E-Fabre.com/en/.

4. Use this poem to **talk about how Fabre taught his children the skills of a natural scientist** *("Observe," Papa says. "Patience shows all!")*. He encouraged observation, listening, and measurement. Discuss each of those skills and how they fit into conducting outdoor investigations.

5. For a poem about observing birds, look for **"Discovery / Descubrimiento" by Margarita Engle** (2nd Grade, Week 4, page 112). And for more poetry about insects, seek out *UnBEElievables: Honeybee Poems and Paintings* by Douglas Florian and *Seeds, Bees, Butterflies and More!* by Carole Gerber.

## Week 5: Predictions & Hypotheses

**Third Grade**

### Testing My Hypothesis
by **Leslie Bulion**

I'm testing my hypothesis:
Cats love the color red.
I think this is the reason that
My cat sleeps on my bed.

I set up an experiment to test it:
False or true?
I swap my clean red blanket for my brother's—
night-sky blue.

I feed my cat and scratch her ears.
My tooth-brushing is done.
I'm undercover, snug in blue.
Experiment's begun!

When morning comes,
My furball's at the bottom of my bed.
"No cat hair here," my brother calls.
"She didn't choose the red."

My cat bats at my wiggling toes.
She sniffs them through the sheet.
The next hypothesis I'll test?
My cat loves smelly feet!

### Take 5!

1. **Create a simple poetry prop**—a card that is red on one side and blue on the other—and read this poem aloud, showing the color that corresponds to the poem as you read.

2. Share the poem again and **invite students to say the color words *red* or *blue*** as they pop up in the poem while you read the rest of the poem aloud. Cue them by using your color card poetry prop.

3. Use this poem to **talk about how young people can do their own research at home on things as simple as pet behavior**. Brainstorm possibilities they might want to try.

4. **Talk about how scientists analyze and interpret patterns in data** to construct reasonable explanations based on evidence from investigations—sometimes over and over again for reliable results. How does this cat experiment follow those scientific principles?

5. For another poem about an informal investigation, look for **"The Class Plant" by Janet Wong** (2nd Grade, Week 5, page 113). And for more poems about cats and cat behavior, seek out *A Curious Collection of Cats* by Betsy Franco and *Won Ton: A Cat Tale Told in Haiku* by Lee Wardlaw.

## WEEK 6: INVESTIGATIONS

### THIRD GRADE

### MEET MR. WIZARD
by **George Ella Lyon**

No science lab in my
school, no library
even. But Mr. Swift
does experiments
anyway.

My favorite is the egg
and the bottle.
The egg has to be boiled
and peeled, the bottle
empty.

You also need a strip
of notebook paper
and a match. Scritch!
Flame flowers at the paper
edge. You drop it in the bottle

and place the egg on
the bottle's lip, blocking
the air so the flame goes
lower
       lower
          out.

There's a pause
and THWUP!
The egg
slooks through the neck.
This demonstrates

the vacuum
and proves
that science
can make
you laugh!

### Take 5!

1. If possible, **the perfect props for this poem would be an egg and a bottle,** as mentioned in the poem (whether you use them to demonstrate or not). Then read the poem aloud, holding up your props.

2. Share the poem again with **students chiming in to highlight the sound words (onomatopoeia) in the poem** *(Scritch!, THWUP!, slooks)* while you read the rest aloud.

3. This poem reminds us that *science can make you laugh!* **Watch a video demo of this experiment here** and note the laughter in the background: YouTube.com/watch?v=ctJyu5ete6Y.

4. Use this poem to **talk about how scientists predict, observe, and record changes in the state of matter caused by heating or cooling.** Heating is an essential part of this investigation. The egg is boiled and peeled; a piece of paper is lit on fire and placed in the bottle, reducing the air pressure inside and creating a vacuum. The higher pressure outside the bottle pushes the egg in. When you blow into the bottle, it makes the air pressure higher inside the bottle, so the egg will come out.

5. Temperature is also an important component in these poems: **"Celsius Thermometer" by Renée M. LaTulippe** (1st Grade, Week 8, page 76) and **"Dog in a Storm" by Stephanie Calmenson** (Kindergarten, Week 17, page 45). For more humorous science poems, look for *BrainJuice: Science, Fresh Squeezed!* by Carol Diggory Shields.

*The Poetry Friday Anthology for Science*

## WEEK 7: DATA

### Take 5!

1. **If you have a penny or a piece of fruit handy, your poetry props are ready.** Show them off while you read this poem aloud.

2. **Next, divide students into two groups**—one to say the names of fruits as they occur in the poem and one to say the statistics for each fruit. Read the poem aloud again, and lift your penny or fruit prop to cue students to participate.

3. Talk with students about data and **how collecting information is an important part of scientific study.** Work together to measure and report data on the spot, such as how many pens, pencils, markers, and other writing utensils you all have among you. Report it in column, table, or graph form.

4. Use this poem to **talk about how energy exists in many forms, including electricity.** Here the poet reveals that *a penny and a galvanized nail / can produce an electrical current in fruit* and tests it on several different examples of fruit, with data included using a voltmeter to measure voltage. Draw a simple conclusion from the data (grapefruits generate more voltage than tomatoes).

5. For another example of energy, look for **"Discovery" by X. J. Kennedy** (1st Grade, Week 5, page 73). And for a nonfiction resource, consult *Battery Science: Make Widgets that Work and Gadgets that Go* by Doug Stillinger.

## ZAPPED!
### by April Halprin Wayland

Dad says
that a penny and a galvanized nail
can produce an electrical current in fruit.

In *fruit?*
Hmm... am I doing this right?
Wait... *WOW!*

Now my heart's racing—
I'm filling pages:
how much voltage did each fruit produce?

Tomato:      .59-.62
Kiwi:        .85-.86
Orange:      .89-.90
Lemon:       .91-.93
Grapefruit:  .93-.94

(Sometimes science makes me hungry.)

I'm so jazzed,
if you measured *my* current
on the voltmeter,

you'd be shocked.

# Week 8: Tools of Science

**Third Grade**

## ARMOR
by **Margarita Engle**

Jagged spikes, prongs, spears, barbs, spines!
Until I peered at tiny pollen grains
under a powerful microscope,
I had no idea that flowers
could be so fierce.

No wonder my nose feels
scratched and scraped
in hay fever season!

**Note:** If you look at a variety of pollen grains, magnified by an electron microscope, you might see protective spikes that keep pollen from being eaten by insects.

### Take 5!

1. Before sharing this poem, **show visuals that reveal pollen under the microscope.** One example features a specialized pollen microscope: YouTube.com/watch?v=FFRU0erMAvM. Then read this poem aloud with a pause between the stanzas.

2. Share the poem again and this time **invite students to utter a big (fake) sneeze at the end of the poem.** You might also want to review how to cover your mouth with the bend in your arms to keep from spreading germs to your hands and to other people.

3. Talk about how **scientists use tools like microscopes to study and analyze information.** If you have a microscope available, review procedures for using one safely. If not, revisit the video link above.

4. **Use this poem to talk about a flower's life cycle.** Investigate the role pollen plays in flower reproduction. Research together why flower pollens might have *jagged spikes, prongs, spears, barbs, spines* and *be so fierce.*

5. For more poems about how organisms adapt to their environments, look for **"What Is a Foot?" by Jane Yolen** (5th Grade, Week 22, page 250) or **"Seeing School" by Kate Coombs** (1st Grade, Week 25, page 93). For more plant-themed poems, look for *I Heard It from Alice Zucchini: Poems About the Garden* by Juanita Havill.

## WEEK 9: MATTER

### THIRD GRADE

### Take 5!

1. **Act like the molecules in a solid, liquid, and gas. Stand straight** like a soldier to be a solid, **then wiggle your arms and feet a little** to be like a liquid, and finally **move your arms and legs in waves** to be a gas.

2. Divide the class **into three groups—one to say each of the "answer" lines:** *"I am," says the wall; "I am," says the milk; "I am! Call me Steam-y."* You read the rest of the poem aloud, pausing at the end of each stanza as you transition to the next item.

3. For discussion: ***What is your favorite state of water—solid, liquid, or gas?*** (Consider eating, drinking, weather, and other scenarios.)

4. Guide students in identifying the ways in which **matter is described and classified by their observable properties in this poem** (a solid's size and shape remain the same; a liquid needs a container to give it shape; gases/vapors fill a room but often can't be seen). Make a chart and list examples of solids, liquids, and gases students are familiar with.

5. Follow up with another poem about liquids, **"Liquids Can't Contain Themselves,"** also by Heidi Bee Roemer (2nd Grade, Week 10, page 118). Also look for Constance Levy's book *Splash! Poems of Our Watery World* or Jane Yolen's *Once Upon Ice and Other Frozen Poems*.

## QUESTIONS THAT MATTER
### by Heidi Bee Roemer

*What is a solid?*

"I am," says the wall.
"My size and shape remain the same;
I don't change at all."

*What is a liquid?*

"I am," says the milk.
"My carton gives me shape.
I'm a puddle when I'm spilt."

*What is a gas?*

"I am! Call me Steam-y!
My vapors fill the room,
but you probably can't see me."

## WEEK 10: MORE MATTER

### THE BRINK
by **Janet Wong**

I fill a cup to the top
with crushed ice,
pour juice to the brim,
neat and nice.
Mom thinks
it's on the brink of disaster.
When I take just a sip,
she shouts: "Drink faster!"
When the ice melts,
will my drink spill out?
I think there's nothing
to worry about
but I wait and I watch.
The ice seems to shrink.
PHEW! Okay—
time to drink!

### Take 5!

1. Add a bit of fun to sharing this poem with a poetry prop—**show a very full glass of ice water before reading the poem aloud.**

2. Read the poem again and **invite students to chime in on the word *PHEW!*** while you read the rest of the poem aloud. Cue them by mopping your brow as if relieved.

3. Check out the U.S. Geological Survey's Water Science School online at Ga.water.usgs.gov/edu/, where **students can test their water knowledge** (in Spanish and Chinese, too!).

4. **Use this poem to talk about how matter has measurable physical properties.** For example, water can be liquid, solid, or gas and has different purposes in each state. Talk about the changes in the water (in the poem) when the cup is full of ice and as the ice gradually melts due to the energy in heat—revealing that any kind of work, such as changing states of matter, requires energy.

5. For another poem with a simple investigation involving temperature changes, revisit **"Meet Mr. Wizard" by George Ella Lyon** (3rd Grade, Week 6, page 154). And for more water-themed poetry, look for *How to Cross a Pond: Poems about Water* by Marilyn Singer.

## Week 11: Force, Motion & Energy

### Take 5!

1. **If possible, bring a brick as a prop to set the stage.** Then read this poem aloud slowly and with pauses between each stanza.

2. For a follow-up reading, **coach students to say the "answer" line, the final line of each stanza.** Read the poem aloud again and pause for their unison line.

3. **Talk about the title of this poem,** "After I Made a Huge Mess With My Chemistry Set," and guess (hypothesize!) what kind of mess that might have been.

4. Use this poem to **focus on how position and motion can be changed by pushing and pulling objects.** Talk about how pushing and pulling a wagon full of bricks was made easier by taking some of the bricks out of the wagon, but that this required more trips. Reducing the load requires less force but more trips to complete the job. If possible, use a spring scale to measure the amount of force in Newtons that it takes to move a brick versus a book.

5. For another poem about pushing a wagon, look for **"Push Power" by Janet Wong** (Kindergarten, Week 11, page 39).

### After I Made a Huge Mess with My Chemistry Set
by **Mary Lee Hahn**

My dad is mad.
He's really ticked.
My punishment?
To move the bricks

he's saving for paving
a path in the garden.
My only tool?
My little red wagon.

I pile it high.
I push and I pull.
My realization?
My wagon's too full.

I take a break
and think it through.
My decision?
Remove just two.

I guess and test
till my load's just right.
My success?
Not too heavy, not too light.

My dad is sad.
He heaves a sigh.
My dreadful chore?
More fun with science!

## WEEK 12: MORE FORCE, MOTION & ENERGY

**THIRD GRADE**

### LIFT
#### by **Marilyn Singer**

When you don't have wings
and feathers, just arms and skin,
gravity will win.

But people, craving
panoramic, learned what's
aerodynamic.

So, with metal, fuel,
invention, victory's
ours: ascension!

### Take 5!

1. Add some fun to sharing this poem with a poetry prop—**show a folded paper airplane before reading the poem aloud.**

2. Read the poem aloud again, and lift your paper airplane prop to cue **students to say the important last word in each stanza** (*win, aerodynamic, ascension*).

3. **Contrast the three forms of energy in this poem:** *wings and feathers* (birds), *arms and skin* (humans), and *metal, fuel, invention* (aircraft). Discuss how each uses energy to move.

4. Use this poem to **talk about another form of energy—mechanical energy—**and the reaction of the air to that energy. This includes four forces: thrust, lift, weight, and drag. Discuss key words and concepts like *aerodynamic, metal, fuel, ascension,* and *gravity* and what role each plays in making flight possible. Helpful explanations are at CT.gov/kids/cwp/view.asp?a=2731&q=330926.

5. Connect this poem with **"Gravity" by Joyce Sidman** (2nd Grade, Week 11, page 119) or **"Space Yacht" by Juanita Havill** (2nd Grade, Week 36, page 144). Look for even more flight-themed poems in *Give Me Wings* edited by Lee Bennett Hopkins and *Make Things Fly: Poems about the Wind* edited by Dorothy Kennedy. And don't miss the informative guide *The Paper Airplane Book* by Seymour Simon.

## Week 13: Light & Sound

### Third Grade

### Sound Waves
by **Michael Salinger**

Sound travels in waves
like ripples from a penny
dropped in a bucket of water.
Once sound is created
a tick, a tock
a clap
a boom
a whisper
a crackle
a crash, a zoom
its energy rides upon the crest
doing what any wave does best.
Traveling through
air, water, or even a solid
until its energy is finally used up
and the noise fades to quiet
the sound flattened to stillness
not making a peep
until that dog next door barks
setting off another wave
waking you from your sleep.

### Take 5!

1. Before sharing this poem, take a moment to **encourage students to close their eyes and listen carefully** to the sounds all around them. After a moment of silence, continue by reading this poem aloud.

2. Next, read the poem aloud again and **invite students to say the sound words** in the poem (*a tick, a tock / a clap / a boom / a whisper / a crackle / a crash, a zoom*) while you read the rest aloud.

3. **Work together to locate examples of the sounds included in the poem** using online resources like SoundCloud.com. Play them while reading the poem aloud together again—and make a recording of the reading.

4. Using the examples in this poem, **talk about how energy exists in many forms, including sound waves.** Consider how the effects can vary with increasing or decreasing amounts of sound energy depending on how close the sound is and how much force is used to make the sound. Rank the examples in the poem from loud to soft. This chart shows common sounds listed according to noise levels: CHEM.Purdue.edu/chemsafety/training/ppetrain/dblevels.htm.

5. Explore a similar poem about sounds we hear every day, **"Sound Waves at Breakfast"** by Susan Marie Swanson (2nd Grade, Week 13, page 121).

## WEEK 14: SPACE

### ORION NEBULA
by **Mary Lee Hahn**

It's huge.
It's far.
The birthplace of
stars.

It's dust.
It's gas.
Gigantic in
mass.

### Take 5!

1. Before reading the poem aloud, **display a star map** that shows an image of the Orion Nebula. Once source is GalaxyMap.org.

2. **The arrangement of lines in this poem lends itself to a "call and response" read aloud** of the poem. Break students into two groups; one will read the first line of each stanza, the other group will read the second line of each stanza while you read the rest of the poem aloud.

3. **Talk with students about how stars look to us from earth** and how we find out about what they actually look like. Use the resources at Space.com/skywatching for visuals, particularly the "Image of the Day."

4. Work with students to identify the planets in Earth's solar system and their position in relation to the sun. **Collaborate to create a quick glog,** a digital interactive poster (using Glogster.com), pulling together images and key words for a visual representation of the planets and sun. Show students the choices of text, fonts, color, graphics, and even animation, if possible, while you input those items and create the finished product.

5. Connect this poem with another about exploring the skies, **"Galileo Galilei" by Renée M. LaTulippe** (3rd Grade, Week 32, page 180), and with selections from *Out of This World: Poems and Facts about Space* by Amy E. Sklansky.

## Week 15: Sun, Earth & Moon

**Third Grade**

### Lunar Eclipse
by **Bobbi Katz**

Like a dark old penny
    being placed
on top of a bright new one,
Earth's shadow slowly slips
over the copper moon
hiding it completely.
Then slowly,
    slowly,
        it slides off
     as if pushed
        by an invisible
            finger.

### Take 5!

1. **Bring two pennies (preferably an old one and a new one) as a poetry prop** to set the stage, then read this poem aloud slowly and dramatically, manipulating the pennies as suggested by the poem.

2. Then read the poem aloud again, and **invite students to chime in on the word *slowly* each time it occurs in the poem** while you read the rest of the poem aloud. A student volunteer can move the pennies this time.

3. **Connect this moon poem with a nonfiction picture book** about moon exploration, *Moonshot: The Flight of Apollo 11* by Brian Floca. Talk about how details are presented in these two different formats.

4. The poem shows how simply two pennies can be used as a model to **demonstrate the relationship of the earth and the moon**, including orbits and positions. Follow up with video demonstrations of a lunar eclipse from sources such as YouTube.com/watch?v=lcRp1jKJmJU.

5. Link this poem with another about the moon, **"Queen of Night" by Terry Webb Harshman** (4th Grade, Week 15, page 203), and with selections from *And Then There Were Eight: Poems about Space* by Laura Purdie Salas; *Comets, Stars, the Moon, and Mars* by Douglas Florian; or *Thirteen Moons on Turtle's Back: A Native American Year of Moons* by Joseph Bruchac.

## WEEK 16: THE WATER CYCLE

**THIRD GRADE**

### WE NEED GREEN SEAWEED!
by **Margarita Engle**

The carbon dioxide/oxygen cycle
is just a circle. Pretty simple.
My nose needs Os,
and green plants crave Cs,
so we take turns recycling
the same airy breeze.

It's easy, as long as no one
makes the ocean dirty, or chops down
all the trees.

### Take 5!

1. To set the stage for this poem, **write the letters O and C on the board**. Then read the poem aloud, emphasizing those two stand-alone letters in particular.

2. Then read the poem aloud again, and this time **invite students to say those letters (O and C) loud and clear**, while you read the rest of the poem aloud.

3. Work with students to **research a few facts about the role the oceans play in providing oxygen** for the entire planet. One helpful website is Kids Do Ecology at Kids.nceas.ucsb.edu/biomes/marine.html.

4. **Use this poem to talk about the relationship between people and plants** (giving and receiving oxygen and carbon dioxide). Resources like this video can help students conceptualize the oxygen-carbon cycle: YouTube.com/watch?v=xFE9o-c_pKg.

5. Another poem that makes a powerful point about our role in the environment is **"Resources Rule!" by Susan Blackaby** (5th Grade, Week 20, page 248). Also look for selections from *Earthways, Earthwise: Poems on Conservation* edited by Judith Nicholls.

## Week 17: Weather & Climate

### Take 5!

1. **Create a visual background for this poem** with a page from Weather.com. Then read the poem aloud with a brief pause between each couplet stanza.

2. In sharing the poem aloud again, **students can say the temperature numbers in the poem** (*93, 90, 75, 50*) for greater emphasis, while you read the rest of the poem aloud.

3. For discussion: ***How does paying attention to weather reports help?***

4. Use this poem to jumpstart a **discussion of how we observe, measure, record, and compare day-to-day weather changes.** Meteorologists report temperatures in multiple locations and often include information about air temperature, wind direction, and precipitation. Check out local television or radio weather reports for examples.

5. Link this poem with another about the weather, **"Rain Gauge" by Anastasia Suen** (2nd Grade, Week 17, page 125). And for even more weather-themed poems, look for *Sharing the Seasons* edited by Lee Bennett Hopkins; *The Year Comes Round: Haiku Through the Seasons* by Sid Farrar; and *Weather Report* by Jane Yolen.

## This Week's Weather
### by Janet Wong

Monday's temperature was 93.
Way too hot.

Tuesday's high was 90.
Better, but not a lot.

Wednesday it cooled down.
It was 75—just right.

And yesterday, a storm came.
It rained cats and dogs all night!

It rained so much
it was like the sky sent down a river.

It was windy and only 50 degrees—
I felt myself shiver.

How could the weather change so much
in a week (or less)?

This is why we have
weather news, I guess!

## Week 18: Forces of Nature

**Third Grade**

### Tornado!
by **Carole Gerber**

Warm, moist air drifts toward the sky;
gets caught in cold air speeding by.
Vicious, raging rains erupt.
Lightning flashes, wind speeds up
and shapes into a funnel form.
Tornado!
Deadly product of the storm.

### Take 5!

1. **Play a clip of a video of a tornado sighting (without sound)** as the backdrop while you read the poem aloud. A good example is at YouTube.com/watch?v=xCl1u05KD_s.

2. Read the poem aloud again, and **invite students to chime in when the title of the poem appears within the poem** (Tornado!).

3. This is a teachable moment for **talking about emergency preparedness**—in school, at home, and in the community.

4. **Use this poem to talk about environmental changes** such as tornadoes, floods, and droughts. What is the impact on the community? (Everyone is displaced, some temporarily and some permanently.) If possible, look for the tornado lab made by Discovery Kids which includes a simple tornado-making kit and a "Storm Chaser" DVD.

5. For the pet perspective on a scary storm, look for **"Dog in a Storm" by Stephanie Calmenson** (Kindergarten, Week 17, page 45). Or for more weather poems, find *A Crack in the Clouds* by Constance Levy and *Seed Sower, Hat Thrower: Poems about Weather* by Laura Purdie Salas. Seymour Simon's nonfiction picture book *Tornadoes* offers more factual details and photos, too.

## Week 19: Soil & Land

### Third Grade

### Take 5!

1. If possible, **show an image of a trilobite as your backdrop** while you read the poem aloud. One source is FossilsForKids.com.

2. Share the poem again, and this time **invite students to chant the first two words of the poem**—the repeated word and title of the poem, *Tri-lo-bite*—while you read the rest aloud.

3. **Talk with students about how fossils are essential to fossil fuels** like petroleum and natural gas. For guidance, look for EIA.gov/kids/energy.cfm?page=oil_home-basics.

4. **Use this poem to talk about how scientists identify fossils as evidence of past living organisms,** which helps in understanding the nature of environments in the past. For more information about fossil formation, look for this resource: RocksForKids.com/RFK/howrocks.html.

5. Link this poem with others about fossils such as **"Fossil Fuels" by Janet Wong** (2nd Grade, Week 20, page 128) and **"Shen Kuo" also by Janet Wong** (5th Grade, Week 31, page 259).

## TRILOBITE
### by Mary Ann Hoberman

*Tri-lo-bite, tri-lo-bite*, that is my name
My body's three-sectioned; my name is the same.
I lived when the seas covered mountains and plains.
Now I am gone but my fossil remains.

*The Poetry Friday Anthology for Science*

## WEEK 20: NATURAL RESOURCES

**THIRD GRADE**

### WHAT MAKES A TURBINE TURN
by **Steven Withrow**

The formless force
that waggles a flag
and shapes a ghost
from a plastic bag

and levitates
a dragon kite
and wrestles with
the trees at night

is named the same
as that airy motion
which blusters
over field and ocean

and charges up
electric motors
with each revolving
round of rotors.

When next you see
a three-armed beast
who might be facing
north-northeast

don't worry if
you feel thin-skinned.
"It's just my pinwheel,"
says the wind.

**Note:** Wind is not the only thing that makes a turbine turn; there are also solar turbines, gas turbines, steam turbines, and more.

### Take 5!

1. After reading this poem aloud, **introduce the key word *turbine*,** essential to understanding the poem. Show an image of a wind turbine from a source like Windustry.org.

2. Then read the poem aloud again and **invite students to say the line that the wind "speaks"** (*It's just my pinwheel*) in a slow, breathy, "windy" voice, while you read the rest of the poem aloud.

3. **Work together to make a simple pinwheel.** A model and directions can be found at Alliantenergykids.com/wcm/groups/wcm_internet/@int/@aekids/documents/document/mdaw/mdiy/~edisp/022821.pdf.

4. **Use this poem to explore the characteristics of wind power** and talk about how it is useful and can help conserve other resources. One helpful resource can be found at Environment.NationalGeographic.com/environment/global-warming/wind-power-profile/. Or look for *Human Footprint* by Ellen Kirk, which shows the impact even children have on the world and its resources.

5. Match this poem with another that explores wind energy, **"Go Fly a Kite" by Laura Purdie Salas** (2nd Grade, Week 12, page 120), or with a poem featuring another form of alternative energy, **"Solar Power" by Susan Blackaby** (4th Grade, Week 20, page 208). Also look for selections from *Make Things Fly: Poems about the Wind* edited by Dorothy Kennedy.

## Week 21: Ecosystems

### Tide Pool
by **Jane Yolen**

Between two rocks,
A world made small,
The naturalist
Can't count it all.

The population
Does sustain
Itself with neither
Milk nor grain

But pulls its living
From the sea
And from the sun
So naturally

That whelk and slug
And sponge and prawn
And crab and star
Live on and on

Between the rocks,
Between the tides,
Where life so bright
Precariously rides.

### Take 5!

1. After reading this poem aloud, **talk about the sea creatures mentioned in the poem:** *whelk, slug, sponge, prawn, crab, star.* Use web resources such as Animals.NationalGeographic.com to show images.

2. As you share the poem again, **invite students to join you on the fifth stanza,** which features the tide pool creatures, while you read the rest of the poem aloud.

3. Challenge students to work in pairs or trios to **draw a picture for one of the tide pool creatures in the poem,** Then post all the pictures in order corresponding to the poem and read the poem aloud together again.

4. **Use this poem to talk about the ecosystem of the tide pool** and the relationships between the animals, water, rocks, and sun. How do the physical characteristics of this environment support the creatures that live in it? One helpful resource is Seaworld.org/wild-world/ecosystems/info-books/tide-pools/index.htm.

5. Compare this poem with one about another ecosystem, **"Tropical Rain Forest Sky Ponds" by Margarita Engle** (4th Grade, Week 21, page 209). Then connect with the poetry books *Ocean Soup: Tide-Pool Poems* by Stephen Swinburne and *Sea Watch: A Book of Poetry* by Jane Yolen.

**WEEK 22: ADAPTATIONS & TRAITS**

**THIRD GRADE**

## CAMOUFLAGE
by **Margarita Engle**

When swift eagles strike from above,
and fast jaguars leap from below,
the gentle sloth stays alive by being
SOOOO very SLOOOOW
that algae and moss turn his fur as green
as a leafy branch
with no place to go.

No wonder he always hangs upside-down
with a sly grin, like a tricky clown.

### Take 5!

1. **Set the stage for this poem by showing a picture of a sloth.** One resource with an extensive library is Animals.NationalGeographic.com. Then read the poem aloud, slowing down as you go.

2. Next, share the poem again and **invite students to chime in on the important fourth line** (*SOOOO very SLOOOOW*), reading it very, very slowly while you read the rest of the poem aloud.

3. Work together to **research facts about International Sloth Day** on October 20.

4. Use this poem to **talk about the concept of "camouflage"** and how the sloth uses movement (or a lack of it) to hide in plain sight, and how *algae and moss turn his fur as green*, protecting it from predators (like the eagle and the jaguar). Why would green fur help the sloth hide?

5. Connect this poem with another one about how organisms adapt, **"What Is a Foot?" by Jane Yolen** (5th Grade, Week 22, page 250), and look for more informative poems in *Chatter, Sing, Roar, Buzz: Poems about the Rain Forest* by Laura Purdie Salas. For a fun nonfiction book, look for *A Little Book of Sloth* by Lucy Cooke, and check out Eric Carle's *"Slowly, Slowly, Slowly," Said the Sloth*.

*THE POETRY FRIDAY ANTHOLOGY FOR SCIENCE*

# WEEK 23: CYCLES

**THIRD GRADE**

## Take 5!

1. As you read this poem aloud, **show a sun scene in the background.** One great source is Calm.com.

2. Since this poem is also presented in Spanish, **invite a reader and speaker of Spanish on your campus or in your community to read the Spanish version aloud.** Then read the English version aloud again. Record both readings for future reference, if possible.

3. Challenge students to work in pairs or trios to **draw a cartoon or picture for one of the lines in the poem.** Then post all the pictures in order corresponding to the poem, in the style of "this is the house that Jack built."

4. Use this poem to **talk about the cycle of food production** described here. In particular, discuss the role of the sun in providing the food we eat. Point out each step along the way, especially the relationship of seeds in the soil to the finished product, bread.

5. For another food + science poem, look for **"Cool Food for Thought"** by Sara Holbrook (5th Grade, Week 21, page 249), as well as Alma Flor Ada's *Gathering the Sun* and the nonfiction picture book *Earth and Sun* by Bobbie Kalman (available in both English and Spanish).

## SUN-KISSED
by Guadalupe Garcia McCall

It is not just your skin that
feels its warm, loving rays.
The sun loved the dirt
who hugged the sun-kissed seed
who unfurled its sun-kissed leaves
and raised the sun-kissed stalk
who produced the sun-kissed wheat
which became a sun-kissed roll
that was baked in a sun-kissed kitchen
and was slathered with sun-kissed butter
because you loved it too.

## BESADO POR EL SOL
por Guadalupe Garcia McCall

No es solo tu piel la que
siente sus rayos cariñosos y cálidos.
El sol amó a la tierra
que abrazó a la semilla besada por el sol
que desdobló sus hojas besadas por el sol
y cultivó el tallo besado por el sol
produjo el trigo besado por el sol
que se convirtió en panecillo besado por el sol
que fue horneado en una cocina besada por el sol
y fue untado con mantequilla besada por el sol
porque también tú lo amabas.

THE POETRY FRIDAY ANTHOLOGY FOR SCIENCE

# WEEK 24: PATTERNS

## THIRD GRADE

### CITIZEN SCIENTIST
by **Shirley Smith Duke**

Scoop up your sweep net.
Come and join me.
We'll search for ladybugs,
rare ones that hide.

Look for the patterns
you'll see on their backs.
Examine them carefully—
nine spots or two?

We'll pull out the camera,
record what we see,
upload our images
to send right off—

The Lost Ladybug Project
needs citizen scientists
and our neighborhood data
in field notes, too,

that we'll write down
for our every find.
Add where and how,
the date, time, and weather.

We're citizen scientists!
We're working together
with other scientists
all over the world!

### Take 5!

1. As you read this poem aloud, **use gestures to suggest the actions in the poem** (scooping a net, looking around, snapping photographs, walking the neighborhood, writing notes).

2. Read the poem again, and **invite students to chime in when the title of the poem appears within the poem** (*Citizen Scientist; We're citizen scientists!*) while you read the rest of the poem aloud.

3. **Research opportunities to participate in various citizen science projects.** Search ScienceBuddies.org for "citizen scientist" alternatives.

4. Talk about how this poem shows how **regular people can join with scientists to study nature and observe patterns that provide significant data** (e.g., nine spots or two on the backs of ladybugs). Consult LostLadybug.org for more information about this specific project.

5. Connect this poem with another about looking for patterns in nature, **"Rings Not Letters" by Juanita Havill** (2nd Grade, Week 24, page 132), and seek out the informational picture book *Ladybugs* by Gail Gibbons.

## Week 25: Human Body

**THIRD GRADE**

### Take 5!

1. **For a poetry prop, show a peanut product** like peanuts or peanut butter. Then read the poem aloud.

2. Then read the poem aloud again, and **invite students to say the key words *sneeze*, *wheeze*, and *allergies*** while you read the rest of the poem aloud.

3. **Here is the teachable moment for talking about various kinds of allergies people can have.** One helpful resource is the American Academy of Allergy, Asthma, and Immunology. Check out their "Just for Kids" link at AAAAI.org/conditions-and-treatments/just-for-kids.aspx. If possible, invite the school nurse or a local health care professional to talk about how to protect children with food allergies.

4. Use this poem to talk about considerate and safe practices and procedures for handling various allergies (like hand washing, not sharing foods, etc.).

5. For another poem about allergies, **revisit "Armor" by Margarita Engle** (3rd Grade, Week 8, page 156) or look for the informative *Yummy Yum for Everyone: A Children's Allergy Cookbook* by Denise McCabe.

### Protecting My Friend
by **Jacqueline Jules**

Pollen in the spring
makes me sneeze.
But peanuts anytime
make Jillian wheeze.
Even a little bit
can make her sick,
so after I eat
I wash up quick.
And I don't worry
about being rude,
because it's safer
not to share my food.
That's how I
protect and defend
kids with allergies,
like Jillian, my friend.

## WEEK 26: KITCHEN SCIENCE

### BREAKFAST ALCHEMY
by **Mary Quattlebaum**

Science in the kitchen!
My little brother claps.
I let him measure flour
(one cup, to be exact).
Next, a spoon of sugar,
one cup of milk, then—crack!—
a single egg, a dash
of oil, and . . .

Flat! Our flapjacks
didn't fatten. Why?
No baking powder! Oops.
A small mistake that we
can mend; we measure, add,
and try again.

My brother stirs and stirs.
Our oopy-goopy glop
grows smooth—and now
the griddle's really hot.
We pour another batch
and soon our liquid, white-ish
dots begin to warm
and thicken, puff and rise . . .
Look now! Oh, wow!

Solid gold upon our plates:
Super-Yummy Science Cakes!

### Take 5!

1. **If you have a bowl or cup and a spoon handy, use them as your prop** while reading the poem aloud. Stir, stir, stir the empty cup for added sound effects. You may also need to explain the word *alchemy*—mixing metals to make gold (which is impossible), a forerunner of modern chemistry.

2. Talk with students about how **the poet repeats the use of exclamations for emphasis in this poem** *(Science in the kitchen!, crack!, No baking powder!, Look now!, Oh, wow!)*. Then read the poem again and invite students to chime in on those exclamatory words and phrases.

3. For discussion: **What other kinds of science experiments can we do in the kitchen** (with mixing and cooking, for example)?

4. Talk about **what happens when two or more materials are combined and then changes in matter are caused by heating.** Pancake batter is a suspension, which is a fluid with small pieces of flour floating in it. As it heats, it undergoes a chemical change to become a solid. This poem also shows that science involves trial and error as those elements vary.

5. Link this poem with another kitchen experiment poem, **"Sugar Water" by Janet Wong** (1st Grade, Week 10, page 78). And for more recipe poems, look for *Arroz con leche: Rice Pudding* and others by Jorge Argueta.

## WEEK 27: VIDEO TECHNOLOGY

**THIRD GRADE**

### Take 5!

1. **Hold a remote control of some kind in your hand** while reading this poem and reveal it as your poetry prop at the end of the poem.

2. Share this poem again, and this time **invite students to chime in with the "answer" to this riddle poem,** *remote*, while you read the rest of the poem aloud.

3. **Talk about how this poem is like a riddle,** with a question, clues, and details shared before the answer is revealed. Share other science riddle poems from *Scientrickery: Riddles in Science* by J. Patrick Lewis.

4. **Use this poem to talk about technology terminology** (*click, scroll, volume, channels, control, wirelessly, power, machines, remote*). When would a remote control be especially useful (for someone with physical disabilities, for someone place-bound, for multi-tasking, etc.).

5. Look for another poem about handheld technology, **"Questions, Questions" by Ann Whitford Paul** (4th Grade, Week 3, page 191).

## WHAT AM I?
### by **Esther Hershenhorn**

I was born long ago,
in 1956.
Zenith's Dr. Robert Adler gave me my *clicks*.

I fit inside your hand,
I turn ON and OFF and scroll.
I raise and lower volume.
I change channels.
I control.

Wirelessly I work from far away.
I give power to machines.
Did you guess *remote*?
Hurray!

WEEK 28: MACHINES

THIRD GRADE

## FIVE O'CLOCK RUSH
by **F. Isabel Campoy**

Game's in thirty minutes
and father is late.
How about dinner?
he asks in dismay.

I open the fridge
and take what I saved
two pieces of pizza
to heat
in
the microwave.

Only one minute.
Dinner is served!


## PRISAS A LAS CINCO
por **F. Isabel Campoy**

En treinta minutos
empieza el partido.
Papá llegó tarde
—¿Y de cena?—
pregunta afligido.

Abro la nevera,
saco mi sorpresa:
¡dos trozos de pizza!
Rápido las meto
en el microhondas.

Y casi enseguida,
en solo un minuto
¡La cena está servida!

### Take 5!

1. If you have a bell handy, or a glass and a metal spoon, use either to **create a ringing noise just prior to the final line of the poem,** as you read this poem aloud.

2. Read the poem aloud again, and **invite students to chime in on the very last line *(Dinner is served!),*** loudly and with enthusiasm, while you read the rest.

3. For discussion: *What other machines (besides microwave ovens) help us make household tasks faster and easier?*

4. **Use this poem to talk about how machines are used in everyday life.** Make a list of machines used in the classroom, school, library, or at home. Create a simple table showing the function of each, what natural resources are used, and what energy sources are needed.

5. Link this poem with another about a household machine, **"Metal Monster" by X. J. Kennedy** (Kindergarten, Week 28, page 56), or **"Microwave Oven" by Janet Wong** (4th Grade, Week 28, page 216). Or look for microwave recipes for young people in *Cook It in a Cup!: Quick Meals and Treats Kids Can Cook in Silicone Cups* by Julia Myall or online at KitchenDaily.com/microwave-recipes/kid-friendly/.

## Week 29: Building Things

### Third Grade

### Take 5!

1. As you read this poem aloud, **use a videotape of a crane in action as your backdrop.** One possible video can be found at YouTube.com/watch?v=UmWFcAX5N9A.

2. Share the poem again, and this time **invite students to read the two important stand-alone lines** (*those details are the key; is construction art*) while you read the rest aloud.

3. **Work together to research other examples of heavy equipment.** One resource is HeavyEquipment.com.

4. Use this poem to **talk about some of the factors involved in moving large objects** (like a tree), including how position and motion can be changed by pushing and pulling objects—as the crane raises and lowers the boom.

5. For more poems on how things work, look for **"Levers"** (1st Grade, Week 28, page 96) and **"Gears"** both by Michael **Salinger** (2nd Grade, Week 28, page 136). And follow this poem with Dotlich's informative picture book *What Can a Crane Pick Up?* along with the digital trailer that goes with it at YouTube.com/watch?v=MkNoQ6TcVDs.

## The Crane Operator
by **Rebecca Kai Dotlich**

He knows which lever to pull
and how to lift that tree,
he's checked the size and weight;

those details are the key

to doing his job, and doing it right—
he raises and lowers
the boom day and night;

he figures the distance,
he studies the chart—
operating a crane

is construction art.

*The Poetry Friday Anthology for Science*

# Week 30: Science Fair

**Third Grade**

## Science Fair Project
by **Eric Ode**

I thought I'd win a ribbon
and my work would be rewarded.
My research, clearly catalogued,
my variables, recorded;
I proudly set my project
with the other kids' displays—
the vinegar volcanoes
and the cardboard hamster maze.
I waited for my trophy,
feeling confident and grand.
And that's about the time
when things got slightly out of hand.
Now my teacher's looking troubled,
and I bet she's holding grudges.
My project ate my tri-fold
and then seven of the judges.

### Take 5!

1. If possible, **display an award ribbon as a prop** before reading this poem aloud, slowly at first and then faster at the end.

2. Share the poem again and **invite students to chime in on the surprising final two lines** (*My project ate my tri-fold / and then seven of the judges*) while you read the rest of the poem aloud.

3. For discussion: **What can you do when your science project goes wrong?**

4. **Brainstorm possible real science fair project ideas with students,** highlighting the importance of clearly recording data. Consult this video series recommended by the National Science Teachers Association: JPL.NASA.gov/education/sciencefair.

5. Connect this poem with another poem with a surprise twist at the end, **"My Project for the Science Fair" by Kenn Nesbitt** (1st Grade, Week 30, page 98).

## Week 31: Famous Scientists

**Third Grade**

### Considering Copernicus
by **Bobbi Katz**

Copernicus, a sage of old,
did not accept what he was told.
He said, "Earth moves around the sun,"
which seemed absurd to everyone,
who saw just how such things were done:
The earth is circled by the sun!
Astronomers and other folk
thought Nicholas was, at best, a joke!
"Sun rises in the east by dawn,
plows through the sky and then is gone
beyond horizons in the west.
Then night
    arrives
        and people rest.
The sun keeps going while we're sleeping,
rising when the birds start peeping."

The process was so obvious
to ALL—
    except...
        Copernicus.

### Take 5!

1. **Display an image of Copernicus in the background** while you read this poem aloud. Then share a short (two-minute) biographical video—both can be found at Biography.com (Biography.com/people/nicolaus-copernicus-9256984).

2. Follow up with another reading of the poem, but this time **stop at the end and let students say the last two words** (*except...Copernicus*).

3. Being an astronomer is one career possibility in science. **Talk about some other science careers** that might interest students. For real life examples, check out this helpful website: SmithsonianEducation.org/Scientist/.

4. Use the details in this poem to **talk about the history of science, what people used to think** (*The earth is circled by the sun*), and how Copernicus changed our understanding about the relationship between the sun, earth, and moon.

5. Pair this poem with another about the role of the sun, **"The Shadow Grows (and Shrinks, and Grows) by Laura Purdie Salas** (5th Grade, Week 15, page 243), and look for the nonfiction book *I Wonder Why the Sun Rises and Other Questions about Time and Seasons* by Brenda Walpole.

## WEEK 32: MORE FAMOUS SCIENTISTS

**THIRD GRADE**

### GALILEO GALILEI
by **Renée M. LaTulippe**

The stars tell stories
of Galileo Galilei.

A genius with a telescope,
he turned his lens upward
and magnified the moon.

He magnified the moon!
And the planets—
the rings of Saturn
the phases of Venus
the four bright moons of Jupiter.

And craters
and moon mountains
and billions of Milky Way stars.

Billions of stars!
Like Copernicus, he knew
that we spin around
a stationary sun.

Centuries ago,
he turned his lens upward
and magnified a universe
of knowledge.

### Take 5!

1. **Display a star map or sky scene in the background while you read this poem aloud.** One source is Space.com and their "Image of the Day." Then point out that Galileo made his discoveries in the 1600s in Italy as the first person to use the telescope to study the sky.

2. Next, share the poem again, and **invite students to say the lines detailing some of Galileo's major discoveries** (*He magnified the moon! / And the planets— / the rings of Saturn / the phases of Venus / the four bright moons of Jupiter. / And craters, / and moon mountains / and billions of Milky Way stars*), while you read the rest of the poem aloud.

3. **Work together to research more facts about Galileo's contributions to science** with the Galileo Project about his life and work at Galileo.Rice.edu.

4. Use this poem to talk about how **scientists rely on empirical evidence, logical reasoning, and experimental and observational testing** to draw conclusions about the world around us. In particular, Galileo studied the moon, its orbit, and its position through the lens of his 17th-century telescope.

5. Revisit the previous poem **"Considering Copernicus" by Bobbi Katz** (3rd Grade, Week 31, page 179). Also look for the poetry book *Galileo's Universe* by J. Patrick Lewis and the biographical picture book *Starry Messenger* by Peter Sís.

# WEEK 33: COMPUTERS

## WIKI ALERT
### by Debbie Levy

Oh, those tricky wikis!
Rounding up wisdom from hither and yon,
Hither says *pro*,
Yon argues *con*,
A tower of facts—
An info-nomenon.

But what is real?
What is true?
Is the truth up to you?
'Midst the cons and the pros
Is there someone who knows?
Could be. I suppose.

To tease out the truth in a wiki
You've got to be wary and picky.

## Take 5!

1. After reading this poem aloud, **show a screenshot from a Wikipedia page (Wikipedia.org)** and talk about how this extensive web resource can be edited by **anyone** and is very useful, if not always correct or reliable.

2. **Next, divide students into two groups**—one to say the words *pro* or *pros* as they occur in the poem and one to say *con* or *cons* while you read the rest of the poem aloud.

3. There are many ways to participate in a local science fair. **Discuss possible *digital* science fair projects** students might consider. Look at ScienceBuddies.org for more information and examples at ScienceBuddies.org/science-fair-projects/Intro-Digital-Photography.shtml.

4. Use this poem to **talk about the importance of evaluating the content of the digital resources we use.** How can we validate and evaluate the relevance and appropriateness of information we find on the Web? (Use multiple sources, consider the source reputation, etc.) Talk about Wikipedia and how it can be changed by anyone.

5. Revisit the poem **"Inquiry" by Cynthia Cotten** (3rd Grade, Week 3, page 151) and read **"Scientific Inquiry" by Susan Blackaby** (5th Grade, Week 1, page 229) to reinforce the importance of questioning in science.

## WEEK 34: SCIENCE CAREERS

**THIRD GRADE**

### INVENTION INTENTIONS
by **Kristy Dempsey**

Welcome to my interview.
I have some questions (just a few),
designed to help me understand
which products might be in demand.
I'm striving for development
of something you'll find relevant:
what do you need, what do you wish
would make your life more simple-ish?
What products do you use the most?
Would you buy pre-buttered toast?
What's more important: cost or ease?
What would you pay me, if you please?
Is there an item you can't stand?
I'll redesign it with my plan!
My product just might fill a niche
Or maybe it would scratch that itch
The one you just can't reach in back . . .
My invention might be what you lack!

### Take 5!

1. Add some fun to sharing this poem with a poetry prop—**hold a clipboard or notepad and a pen** as if you're interviewing someone while you read this poem aloud.

2. Invite students to select their favorite question line and to **chime in on that line only** while you read the rest of the poem aloud.

3. **Use the question lines in this poem as a prompt for class discussion.** In particular, *what do you need, what do you wish / would make your life more simple-ish? / What products do you use the most? / Is there an item you can't stand?*

4. Much of **scientific investigation and discovery is based on asking and answering questions**, making inferences, and selecting and using equipment or technology to solve a specific problem. Model this process by choosing one of the items from the brainstorming activity above and identifying the steps toward a solution or invention. Look for help at KidsInvent.org.

5. For more poems about inventions and innovations, look for **"Printing, Pressed Beyond Words . . . " by Robyn Hood Black** (5th Grade, Week 33, page 261) and **"The 'Black Leonardo'" by J. Patrick Lewis** (1st Grade, Week 32, page 100). Also seek out selections from *Incredible Inventions* edited by Lee Bennett Hopkins and *A Burst of Firsts* by J. Patrick Lewis.

## Week 35: Future Challenges

### Cancer
by **Mary Lee Hahn**

Cancer's *what*
is cells growing wild.

Cancer's *who*
is man, woman, or child.

Cancer's *why*
is scientists aren't sure.

Cancer's *hope*
is someday a cure.

Cancer's *enemies*
are surgery and drugs.

Cancer's *helpers*
are flowers and hugs.

### Take 5!

1. Before reading this poem in a quiet voice, point out to students that **many poems are funny, but some are serious**—like this one.

2. Share the poem again, and **invite students to chime in on the words in italics** (*what, who, why, hope, enemies, helpers*) while you read the rest of the poem aloud.

3. Talk with students about **what we do know about preventing cancer** (don't smoke, eat a balanced diet, get regular exercise, sleep enough, etc.).

4. Use this poem to **focus on the key questions raised for tackling the problem of cancer** (*what* is it, *who* gets it, *why* do people get it). Consider the role scientists play in testing and experimenting as they search for cures.

5. Another poem that addresses the question of why people get sick is **"I Want to Know Why" by David L. Harrison** (1st Grade, Week 35, page 103). Also look for poems written by children with cancer such as *Heartsongs* by Mattie Stepanek.

## WEEK 36: FUTURE DREAMS

### MOVING TO ATLANTIS CITY, 2112
by **Steven Withrow**

Eleven billion side by side
Take up a lot of space.
We needed fresh solutions quick
To house the human race.
We first built floating towns as big
As islands on the sea.
But soon these grew too overstuffed
For Mom and Dad and me.
We'd watched a holo on the web:
*New Lab Needs Volunteers!*
Our family signed up to be
Among the pioneers.
Of all the kids they picked to live
Beneath the ocean's rim
I wasn't quite the perfect choice.
I couldn't even swim.
But once we rode the shuttle down
And passed the pressure sphere
I knew there was no surface place
So much like Home as here.
Although I sometimes miss the sun
And unrecycled air,
I'd never trade my deep dark view
With all of you up there.

### Take 5!

1. **Display an image of an underwater landscape in the background** while you read this poem aloud. Search "landscapes" at Photography.NationalGeographic.com.

2. **Invite students to make a simple digital sign** (on a computer or tablet) saying *New Lab Needs Volunteers!* Then they can say that line of the poem while you read the rest aloud.

3. **Talk about the title of this poem,** "Moving to Atlantis City, 2112," and the legend of Atlantis, a lost island city that sank in the ocean. Why did the poet choose this title?

4. This poem describes an imaginary underwater home, but it's not that far from reality. Share the National Geographic video about "Utopian Underwater Living" at YouTube.com/watch?v=iKXVX3fkxWE. **Discuss the pros and cons of undersea living and how we might tackle the problems** creatively in the future.

5. Pair this poem with another about a real-life ocean scientist, **"Ocean Explorer Sylvia Earle" by Leslie Bulion** (Kindergarten, Week 32, page 60), or revisit **"We Need Green Seaweed!" by Margarita Engle** (3rd Grade, Week 16, page 164). Look for more ocean-themed poems in *Water Sings Blue: Ocean Poems* by Kate Coombs, as well as the picture book biography *The Fantastic Undersea Life of Jacques Cousteau* by Dan Yaccarino.

# Poems for Fourth Grade

# NGSS Science and Engineering Practices: Fourth Grade

*These practices form the foundation of disciplinary literacy in science and integrate reading, writing, listening, and speaking skills from the language arts. Here we indicate which weekly poems emphasize which science and engineering practices at each grade level.*

| PRACTICE | POEM |
| --- | --- |
| Asking questions and defining problems | Weeks 1, 3, 16, 30, 32 |
| Developing and using models | Weeks 4, 8, 27, 36 |
| Planning and carrying out investigations | Weeks 5, 28, 29 |
| Analyzing and interpreting data | Weeks 6, 10, 12, 15 |
| Using mathematics and computational thinking | Weeks 7, 25 |
| Constructing explanations and designing solutions | Weeks 9, 14, 18, 20, 22, 33, 35 |
| Engaging in argument from evidence | Weeks 2, 19, 31 |
| Obtaining, evaluating, and communicating information | Weeks 1, 11, 13, 17, 21, 23, 24, 26, 34 |

# Fourth Grade

| | | |
|---|---|---|
| week 1 | Scientific Practices | What Is Science? *by Cynthia Cotten* |
| week 2 | Lab Safety | Dinos in the Laboratory *by Kristy Dempsey* |
| week 3 | Ask and Ask Again | Questions, Questions *by Ann Whitford Paul* |
| week 4 | Observations | A Biological Community/Una comunidad biológica *by Margarita Engle* |
| week 5 | Predictions & Hypotheses | A Dog's Hypothesis: Zoey's Guide . . . *by Susan Taylor Brown* |
| week 6 | Investigations | Froggy *by Charles Waters* |
| week 7 | Data | Nursing Math *by Jeannine Atkins* |
| week 8 | Tools of Science | Computer Models *by Janet Wong* |
| week 9 | Matter | Changes *by Janet Wong* |
| week 10 | More Matter | What Can You Make from Carbon? *by Laura Purdie Salas* |
| week 11 | Force, Motion & Energy | Roller Coaster Ride *by Patricia Hubbell* |
| week 12 | More FM&E | Friction *by Sara Holbrook* |
| week 13 | Light & Sound | To the Eye *by Laura Purdie Salas* |
| week 14 | Space | The NEO Hunters *by Juanita Havill* |
| week 15 | Sun, Earth & Moon | Queen of Night *by Terry Webb Harshman* |
| week 16 | The Water Cycle | Oh Water, My Friend/Ay agua, mi amiga *by Guadalupe Garcia McCall* |
| week 17 | Weather & Climate | Weather Map *by Joan Bransfield Graham* |
| week 18 | Forces of Nature | Hurricane Hideout *by Janet Wong* |
| week 19 | Soil & Land | What We Eat *by Joseph Bruchac* |
| week 20 | Natural Resources | Solar Power *by Susan Blackaby* |
| week 21 | Ecosystems | Tropical Rain Forest Sky Ponds *by Margarita Engle* |
| week 22 | Adaptations & Traits | Grafting *by Janet Wong* |
| week 23 | Cycles | Windfall in The Andrews Forest *by Joseph Bruchac* |
| week 24 | Patterns | Cicada Magic *by Heidi Mordhorst* |
| week 25 | Human Body | Moving for Five Minutes Straight *by Betsy Franco* |
| week 26 | Kitchen Science | Food for Thought *by Robyn Hood Black* |
| week 27 | Video Technology | Virtual Adventure *by Renée M. LaTulippe* |
| week 28 | Machines | Microwave Oven *by Janet Wong* |
| week 29 | Building Things | Foundation (Don't Rush It!) *by Charles Waters* |
| week 30 | Science Fair | Science Fair *by Irene Latham* |
| week 31 | Famous Scientists | Albert Einstein *by Julie Larios* |
| week 32 | More Famous Scientists | Hawking Time *by James Carter* |
| week 33 | Computers | My Photo Experiment *by Janet Wong* |
| week 34 | Science Careers | Rocky Rescue *by Robyn Hood Black* |
| week 35 | Future Challenges | Shade-Grown *by Margarita Engle* |
| week 36 | Future Dreams | The Real Thing *by Linda Sue Park* |

The Poetry Friday Anthology for Science

*"**Poems and poetry are, for me, a deep form of knowing**, just like science. Yes, obviously, they are different. But each, in its way, is a way to understand the world."*

— Adam Frank

## Week 1: Scientific Practices

**Fourth Grade**

### Take 5!

1. Before sharing this poem, ask students to **listen for key words that describe what "science" is.** Read the poem aloud, and then work together to compile a word wall of terms that describe what *Science is*. Examples include *knowledge, process, experimentation, observation, results, discoveries, questioning,* etc.

2. In a follow-up reading, **invite students to join you on the important opening line (*Science is*) as it occurs repeatedly** while you read the rest of the poem aloud.

3. **Work together to research and identify the various disciplines that are included in the field of science** (physics, astronomy, chemistry, biology, geology, etc.). Talk about the many possible science careers listed at ScienceBuddies.org.

4. **Talk with students about how all scientists rely on asking questions and thinking critically in their work.** They wonder, ask questions, follow a process, conduct investigations and experiments, observe and note, collect data, and analyze and reason logically. How can these steps be useful in other ways, too?

5. Connect this poem with another about **"How to Be a Scientist" by Amy Ludwig VanDerwater** (1st Grade, Week 1, page 69). Also look for books in the "Scientists in the Field" series by Sy Montgomery, such as *The Tarantula Scientist*.

## What Is Science?
**by Cynthia Cotten**

Science is
knowledge
and
a process

Science is
how and why (or why not)
experimentation and observation
expected results and accidental discoveries

Science is
questioning
exploring
understanding

Science is
useful
exciting
ongoing
challenging

Science is
the study of
elements, compounds, cells
motion, sound, light
and of
the ocean,
the earth,
the sky,
and all that they contain

With all that science *is*
perhaps the question should be
"What *isn't* science?"

## WEEK 2: LAB SAFETY

**FOURTH GRADE**

### DINOS IN THE LABORATORY
by **Kristy Dempsey**

I present to you, my esteemed guest,
a theory putting all else to rest,
a scientific extinction story:
Dinos in the Laboratory.
As archeologists dig down down down,
some curious items have NEVER been found,
perhaps providing us a clue
why dinos paid their mortal due:

√     No safety glasses on their eyes
√     No gloves or aprons, dino-size
√     No rules displayed on any wall
√     No regulations to follow at all
√     No fire extinguisher on hand
√     No first aid kit for quick demand

It's plain to see in science class,
these dinos surely DID NOT PASS!
Did science make them go extinct?
I might be wrong. What do you think?
If you have your own suspicions,
don't recreate these dino conditions.
Take my advice. Expect the worst.
Always remember, safety first!

### Take 5!

**1. To set the stage for this poem, show images of dinosaurs** from the paleobiology link at the Smithsonian: Paleobiology.si.edu/dinosaurs/. Then read the poem aloud, pausing dramatically before and after the bulleted list.

2. Share the poem again, and **invite students to chime in on the bulleted list of lines beginning with *No*** while you read the rest aloud.

3. **Use this poem as a prompt to talk about actual theories about dinosaur extinction.** One resource is Library.ThinkQuest.org/C005824/extinction.html.

4. **When it comes to science, it's important to begin by talking about science safety procedures.** Research together what your school, district, or community mandates for classroom and outdoor investigations. This might include wearing safety goggles, washing hands, using materials and equipment appropriately, etc. Then create a simple poster highlighting these guidelines—with or without a dinosaur image!

5. Link this poem with another about lab safety, **"Things to Do in Science Class" by Laura Purdie Salas** (3rd Grade, Week 2, page 150). Just for fun, seek out Robert Weinstock's collection of dino poems, *Can You Dig It?*

## Week 3: Ask and Ask Again

### Fourth Grade

## Questions, Questions
by **Ann Whitford Paul**

How does this phone,
smaller than my hand,
thin as a candy bar,
carry conversations
through space
back and forth
and back again?
Why, with miles between us
and no tube,
no wire, connecting us
we can still hear
loud and clear?
How is it possible
we can talk
for hours . . . hours
without other conversations
jumbling into ours?
Questions, questions!
Where are the answers?
I need to find them all.

### Take 5!

1. Add a bit of fun to sharing this poem with a poetry prop—**show a cell phone before reading the poem aloud.**

2. Read the poem aloud again and **invite students to chime in when the title of the poem appears within the poem** (*Questions, Questions*). Cue them by cupping your hand behind your ear.

3. This poem asks questions about how cell phones work. **Invite students to raise questions about other everyday objects and how they function** (e.g., computers, cars, etc.). Then look up some quick answers at HowStuffWorks.com.

4. Use this poem to talk about the importance of asking questions in science. **Consider how scientists plan and implement investigations by asking well-defined questions**, formulating a hypothesis or making a claim based on evidence, and selecting and using appropriate equipment or technology to answer those questions.

5. Connect this poem with another about a technological invention we use every day, **"What Am I?" by Esther Hershenhorn** (3rd Grade, Week 27, page 175), or share selections from *Incredible Inventions* edited by Lee Bennett Hopkins.

## WEEK 4: OBSERVATIONS

**FOURTH GRADE**

### A BIOLOGICAL COMMUNITY/ UNA COMUNIDAD BIÓLOGICA
by/por **Margarita Engle**

Students trade treats at lunchtime.
*Los estudiantes intercambian bocaditos.*

Ants on the soccer field discover cookie crumbs.
*Las hormigas descubren en el campo de fútbol migajas de galletas.*

Sparrows under the tables find plenty of food.
*Los gorrioncitos encuentran mucha comida debajo de las mesas.*

Flies in the trash cans eat what no one else wants.
*Las moscas de los basureros comen lo que nadie quiere.*

Every organism has its own niche.
*Cada organismo tiene su lugar propio.*

### *Take 5!*

1. Start this poem with a **poetry prop—a lunch bag or lunch box—** placed in front of you. Then read it aloud with a brief pause before each line.

2. Since this is a bilingual poem, it's the perfect opportunity to **invite anyone who speaks Spanish (in your class, school, or community) to join you by reading the Spanish lines** after you read each line in English. You could use VoiceThread to record the Spanish reading with a Spanish speaker in advance.

3. For discussion: **What is your niche (role or job) either at school or at home?**

4. In this poem, we look at how various creatures—including children—look for food in their environments. **Ask students if they have observed any of the things noted in the poem** (students trading treats, ants discovering crumbs, and so on). **Talk about the descriptive word, *niche*, and what it means** —one's place in a habitat as well as one's role or job. Look for more examples in your community.

5. Connect to another poem about outdoor observations, **"Classroom in the Meadow" by Jeannine Atkins** (3rd Grade, Week 4, page 152). Or look for Jorge Argueta's *Tamalitos: Un poema para cocinar/A Cooking Poem.*

# Week 5: Predictions & Hypotheses

**Fourth Grade**

## Take 5!

1. **What is the perfect poetry prop for this poem? A dog biscuit, of course!** If you don't have one handy, cut a few simple bone shapes out of construction paper as your props and wave them at the end of each stanza, dropping them during the final line.

2. **Next, divide students into two groups**—one to say the word *if* as it occurs in the poem and one to say *then* each time it appears in the poem. Cue them by raising your index finger high each time.

3. **Talk about why various dog behaviors might get different results** in the quest for a dog treat (e.g., jumping can be annoying, fetching can be helpful).

4. Use this poem to talk about how the dog's behavior tests a "hypothesis" or claim. For example, *race to the door and bark and jump* results in no treat; *staring at the cookie jar with sad eyes* results in one treat; *fetch a toy when she asks* (and more) results in many treats. **Discuss how scientists perform repeated investigations and identify questions, collect and analyze data, and interpret patterns** to construct reasonable explanations that can be observed and measured.

5. For another poem about a dog and his reactions, share **"Dog in a Storm" by Stephanie Calmenson** (Kindergarten, Week 17, page 45) or look for more poetry about dogs such as Douglas Florian's *Bow Wow Meow Meow* or Betsy Franco's *A Dazzling Display of Dogs*.

## A Dog's Hypothesis: Zoey's Guide to Getting More Goodies
### by Susan Taylor Brown

If
when my human comes home,
I race to the door and bark and jump
to let her know how happy I am
that she came home to me—
then
I will get a goodie.

Result?
A pat on the head
and not one treat to eat.

If
I stand next to her,
staring at the cookie jar with sad eyes,
letting her know I am starving, wasting away,
letting her know I will soon be an invisible dog—
then
I will get a goodie.

Result?
One tiny treat,
barely enough to taste.

If
I whine, just a little,
then fetch a toy when she asks—
if I sit and lie down,
if I even play dead
(I am good at pretending)—
then—

Result?
Jackpot!
Little liver cookies
fall on the floor
all around me.

## WEEK 6: INVESTIGATIONS

**FOURTH GRADE**

### FROGGY
by **Charles Waters**

Sleeves rolled up, utensils at the ready,
I look down at who we're about to dissect
for science class.
I take a deep breath as
Froggy is sliced open
like a hunk of cheese.
Some kids walk away from this spectacle.
I'm fascinated:
*kidneys, lungs, intestines,*
*gall bladder, pancreas, heart,* and *liver*
are all connected like a symphony.
"You have these same parts within you,"
Mrs. Lance says.
"When it comes to our animal friends,
we're more alike than we are different."
"Wait a minute," I say,
looking into my magnifying glass.
"There's one thing
we don't have in common with Froggy."
"What's that?" Mrs. Lance asks.
"Our last meal before we die won't be
beetles, ladybugs, earwigs, and slugs."

### Take 5!

1. As a backdrop for this poem, **show an image of a frog dissection** (frog-life-cycle.com/diagram-frog-anatomy.html) or use a frog dissection app (frogvirtualdissection.com). Then read the poem aloud slowly.

2. Share the poem again and **invite students to chime in on the list of frog body parts** (*kidneys, lungs, liver, intestines, gall bladder, pancreas, heart* and *liver*) **as well as the list of last meal ingredients** (*beetles, ladybugs, earwigs, and slugs*) while you read the rest of the poem aloud.

3. Using the frog diagram or app, identify the location of each of the frog's body parts listed in the poem. Then **compare that to where those same organs are located in our own human bodies.**

4. This poem serves as one example of how scientists carry out descriptive investigations, collecting and recording data by observing and measuring, using tools like hand lenses, and using descriptive words and labeled drawings. **Identify each of these steps together using examples from the poem.**

5. Compare this poem to another that describes an investigation, **"Meet Mr. Wizard" by George Ella Lyon** (3rd Grade, Week 6, page 154), and look for more poems about frogs in *Lizard, Frogs, and Polliwogs* by Douglas Florian.

## WEEK 7: DATA

### Take 5!

1. **Before reading this poem aloud, hold up a pie chart diagram** and ask students to guess what this poem is about.

2. Share the poem again and **invite students to say the exclamation phrase** in the second stanza (*She'd calculate!*) while you read the rest of the poem aloud.

3. **Do a bit of quick collaborative research** on the life and work of Florence Nightingale.

4. This poem shows how science (and math and statistics) can be valuable in multiple careers. Talk about how Nightingale's initiative in collecting and analyzing data to interpret patterns and construct reasonable explanations helped shape medical care for her patients. **Consider the value of tables, charts, bar graphs, line graphs, and pie charts to organize, examine, and evaluate data.**

5. For another poem that emphasizes the value of diagrams, look for **"Let Me Join You" by Heidi Bee Roemer** (5th Grade, Week 25, page 253). Read about other groundbreakers in *Eureka! Poems about Inventors* by Joyce Sidman.

## NURSING MATH
by **Jeannine Atkins**

Florence Nightingale measured medicine and bandaged
arms and legs. She drew charts showing who got strong
and which soldiers stayed sick too long. She saw that
hospitals must be made cleaner and make other reforms.

How could she make this clear to Queen Victoria,
whose eyes glazed at numbers and words? She'd calculate!
The good nurse studied statistics, then drew columns
and charts, with lines curvy and straight.

At the castle, she curtseyed, then showed a circle
divided by lines and with labels. The queen found
it easy as pie to see why change must come, and it would,
from a mathematical nurse and a woman wearing a crown.

**Note:** Florence Nightingale (1820-1910) is famous as the founder of modern nursing. She also developed ways to use diagrams and graphs, creating a forerunner of what is now known as a pie chart.

## WEEK 8: TOOLS OF SCIENCE

**FOURTH GRADE**

**COMPUTER MODELS**
by **Janet Wong**

We have an engineer
visiting our classroom.
She shows us
how she uses her computer
to test designs:
Plan A,
Plan B,
Plan C.
She doesn't
have to build
a whole building
to see if it's better
to have six floors or seven or eight.
She can calculate
how much steel and glass
each different plan
would need,
how much heat
it would use and lose,
how many hours
of construction time
and the cost.
She can tell us everything
about the plans, and—
*click-click-click-click-click!*—
with her camera
and five minutes
of cut and paste
and Photoshop,
she can put us
inside her buildings,
waving hello!

### Take 5!

1. **As a backdrop for this poem, display a computer model of a building**. One resource is ArchitectStudio3d.org/AS3d/for_teachers.html. Then read the poem aloud, stopping briefly wherever punctuation suggests a pause.

2. Share the poem again, and **invite students to join in on the words, *Plan A, Plan B, Plan C,* and then *click-click-click-click-click!*** while you read the rest of the poem aloud.

3. **Talk with students about the different things that engineers do**. One helpful resource is DiscoverE.org, particularly this link: Discovere.org/discover-engineering/engineering-careers.

4. Use this poem to discuss the tools that engineers and scientists might use in their work, including computers and cameras. **Talk about the role of simulations in the design phase and what the pros and cons might be** (save time and money, but cannot anticipate every physical obstacle).

5. Link this poem with another about building, **"The Great Pyramid of Giza" by Laura Purdie Salas** (5th Grade, Week 29, page 257), or for a look at a building across the ages, share *The House* by J. Patrick Lewis.

## Week 9: Matter

**FOURTH GRADE**

## Changes
by **Janet Wong**

Physical change:
your popsicle starts to melt
in the sun.
Pop the popsicle
back in the freezer.
The change can be undone.

Chemical change:
frying a burger
you burn the bottom black.
With a chemical change
unfortunately
there's no going back.

### Take 5!

1. Before sharing this poem, crumple a sheet of paper and **ask students to guess whether you have caused a physical change in the paper or a chemical change** (it's a physical change). Then read the poem aloud, pausing between the two stanzas.

2. Share the poem again, and **encourage students to say first line of each stanza** (*Physical change; Chemical change*) while you read the rest of the poem aloud.

3. **Work together to create a simple chart with two columns: physical change and chemical change.** Then research, list, and discuss examples of each. One resource is Chem4kids.com.

4. **This poem reminds us that all matter has measurable physical properties** and those properties determine how matter is classified, changed, and used. Talk with students about how we measure, compare, and contrast physical properties of matter, including size, mass, volume, states (solid, liquid, gas), temperature, magnetism, color or lack of it, and the ability to sink or float.

5. Connect this poem with another about the properties of common everyday objects, **"Imagine Small" by Eileen Spinelli** (2nd Grade, Week 9, page 117), or look for the *DK Eyewitness Book: Chemistry* by Ann Newmark for more examples.

## Week 10: More Matter

### What Can You Make from Carbon?
by **Laura Purdie Salas**

Charcoal
Makes drawings and fire

Graphite
Makes words that inspire

Diamond
Makes drill bits and rings

Carbon
Makes all living things

### Take 5!

1. Add a bit of fun to sharing this poem with a poetry prop—**show a pencil before reading the poem aloud.** Then point out that the graphite in the pencil is what makes marks on the page.

2. Share the poem again, and **invite students to say the one-word line that introduces each couplet** while you read the rest of the poem aloud.

3. Invite students to choose one thing to quick-sketch from this poem. **Create a quick collage of their sketches arranged around a copy of the poem** to show the variety of possible poem interpretations.

4. Use this poem to talk about how all matter has measurable physical properties. **Work together to research a quick list of things made from elemental carbon**—like the charcoal, graphite, and diamonds mentioned in the poem (hint: *Carbon / Makes all living things*). Discuss the word "organic," which relates to anything with carbon in it, like *organic chemistry*, and compare it with *organic farming*, which doesn't use artificial products or pesticides to help animals and produce grow.

5. Compare this poem to **"Questions That Matter" by Heidi Bee Roemer** (3rd Grade, Week 9, page 157), which addresses various states of matter. For more examples, seek out *Central Heating: Poems about Fire and Warmth* by Marilyn Singer.

## Week 11: Force, Motion & Energy

**Fourth Grade**

### Roller Coaster Ride
by **Patricia Hubbell**

We're fastened in and up we go—
To reach the top we're starting slow—

Whooooshhh!

We're diving! Veering!
Climbing! Swooping!
Dropping! Twisting!
Curving! Looping!

Up! Down! Around! Around!
Down! Down! Down!
We're zooming fast! We're racing faster!
Faster! Faster! Faster! Faster!

We're yelling! Screaming!
Shouting! Laughing!
Hooting! Hollering!
Shrieking! Gasping!

What a crazy, great sensation—
All because of . . . ACCELERATION!

### Take 5!

1. What is the perfect backdrop for this poem? A video of a roller coaster ride! **Play this video with sound effects while you read the poem aloud expressively:** YouTube.com/watch?v=4-5KYPHjp6Y

2. Play the video and read the poem aloud again, but invite students to add additional sound effects to accompany the poem (*yelling, screaming, shouting, laughing, hooting, hollering, shrieking, gasping*) and everyone joins in on the final word, *"Acceleration!"*

3. Challenge students to work in pairs or trios to **draw a picture for one stanza of the poem** then post all the pictures in order corresponding to the poem.

4. **This poem offers a real life example that shows the effect of force on an object** such as a push or a pull, gravity, friction, or magnetism. Work together to research how roller coasters work using sources such as Science.HowStuffWorks.com

5. Pair this with another poem about a fast-moving car, **"Designing an Experiment" by Avis Harley** (5th Grade, Week 6, page 234), and look for *American Coasters: A Thrilling Photographic Ride* by Thomas Crymes or *Roller Coaster* by Marla Frazee.

## WEEK 12: MORE FORCE, MOTION & ENERGY

**FOURTH GRADE**

### FRICTION
by **Sara Holbrook**

Speed bumps in the parking lot.
Gravel under my wheel.
Brakes on a subway train
screeching out a squeal.
A zombie dragging a ball and chain.
My eraser tearing at paper.
My father's weekend beard
on Monday pulling at his razor.
A thumb against a finger
when it
makes a snapping sound.
Whatever takes off in a hurry,
friction slows it down.

### Take 5!

1. Read this poem aloud and **act it out with a snap of your fingers at the end.**

2. Then read the poem aloud again, and **invite students to chime in on their favorite example of friction, saying those lines only** (*speed bumps, gravel, brakes, zombie, eraser, beard, finger snapping*) while you read the rest of the poem aloud.

3. **Work together to conduct quick on-the-spot demonstrations of friction** with erasers and school supplies or hands and feet.

4. **Use this poem to discuss the effect of friction on objects.** Friction can create drag, noise, erosion, speed reduction, damage, and so on. Challenge students to cite examples of each. Talk about ways to reduce friction, too, like using oil in automobiles, etc.

5. Compare this poem with another about force and energy, **"After I Made a Huge Mess with My Chemistry Set" by Mary Lee Hahn** (3rd Grade, Week 11, page 159), and look for the *Time for Kids* reader, *Drag! Friction and Resistance* by Stephanie Paris.

## Week 13: Light & Sound

**Fourth Grade**

### Take 5!

1. Before reading this poem, **turn the lights out** (if possible), then turn them on again as you read this poem aloud.

2. Share the poem again, and **invite students to chime in on the word** *eye* each time it appears in the poem. Cue students by pointing to your own eye.

3. Take a moment to have students close their eyes and describe what they hear, then open their eyes and identify what they see. **Work together to make a quick Venn diagram of things seen only, things heard only, and things that can be both seen and heard.**

4. **Use this poem to talk about light as an energy source and how it enables us to see.** Conduct a quick experiment with light (such as making a simple sundial, using a mirror to reflect light) based on resources such as *Tabletop Scientist: The Science of Light* by Steve Parker.

5. Contrast this poem with **"Sound Waves at Breakfast" by Susan Marie Swanson** (2nd Grade, Week 13, page 121) and look for more poems about light in *Flicker Flash* by Joan Bransfield Graham.

## To the Eye
### by Laura Purdie Salas

Light is
a jumpy kid
playing hopscotch

bouncing from thing to thing
picking up color like a pebble
to carry in its fingers

lightbulb      to French fry      to eye

sun      to tire swing      to eye

campfire      to s'mores      to eye

always, always to the eye

When light lands, you

                     don't
                     hurt
                     don't
                     squint
                     don't
                     blink
                     you

see

## Week 14: Space

**Fourth Grade**

### The NEO Hunters
by **Juanita Havill**

We are hunters
who watch the sky
with telescopes
and camera eye.

We watch for signs
of NEOs
and seek to know
where each one goes.

A Near Earth Object,
an asteroid,
as big as a car,
in the cosmic void
is a rock in space
we want to avoid.

### Take 5!

1. **To set the stage for this poem, show a video of the Near Earth Object (NEO), Asteroid 2012 DA14.** One example is at YouTube.com/watch?v=VsBUZy1ZCYQ. Then read the poem aloud in a quiet and dramatic voice.

2. Share the poem again, and **invite students to chime in on the abbreviation, *NEO*, as well as its elaboration, *Near Earth Object*,** while you read the rest of the poem aloud.

3. **Talk with students about the world's largest public space event, World Space Week,** supported by the United Nations. For information, see WorldSpaceWeek.org/wsw/index.php. Make plans to participate in the next celebration in some way.

4. Use this poem to talk with students about the components of our solar system. **Work together to research how asteroids fit within this system and how they are tracked and studied.**

5. Link this poem with another about studying space, **"Comet Hunter" by Holly Thompson** (5th Grade, Week 14, page 242), and with selections from *Comets, Stars, the Moon, and Mars* by Douglas Florian.

## WEEK 15: SUN, EARTH & MOON

**FOURTH GRADE**

### QUEEN OF NIGHT
by **Terry Webb Harshman**

I am the moon, Queen of Night,
riddle wrapped in borrowed light,

a silver spool where dreams unwind,
ancient orb as old as time.

I masquerade; I wax and wane . . .
forever changing yet the same;

I stir the tides with unseen hands;
they ebb and flow from sea to sand.

Father Sun may keep the day;
I ride along the Milky Way . . .

holding court with owls and bats,
moles and voles and backstreet cats.

Within my tent the weary rest;
puppies doze and sparrows nest.

Children dream beneath my light . . .
I am the moon, Queen of Night.

---

### Take 5!

1. Before sharing this poem, alert students to listen for particular "moon vocabulary." **Then read the poem aloud and make a list of all the moon-specific language they can identify** (e.g., night, light, ancient, orb, wax, wane, tides, ebb, flow, Milky Way).

2. Read the poem aloud again, and **invite students to chime in on the first and last lines of the poem** (*I am the moon, Queen of Night*), echoing the title of the poem).

3. **Connect this poem with a nonfiction book about the moon to compare the factual information you can glean from each source.** One example is *The Moon* by Seymour Simon.

4. **Use this poem to talk about what we know about the moon,** beginning with attributes of the moon and then considering tides, seasons, and the observable appearance of the moon over time during its phases. Consult the NASA website at SolarSystem.NASA.gov/planets/profile.cfm?Object=Moon.

5. Follow up with another poem about the moon, **"I Like that Night Follows Day" by April Halprin Wayland** (1st Grade, Week 24, page 92), and selections from *A Full Moon Is Rising* by Marilyn Singer and *Dark Emperor* by Joyce Sidman.

## WEEK 16: THE WATER CYCLE

**FOURTH GRADE**

### OH WATER, MY FRIEND
by **Guadalupe Garcia McCall**

Are you scared?
Is that why you run
to rivers, to streams, to lakes?
Do you feel safer pooled in ponds?
Does it hurt to boil?
Does your anger roll and roil?
Is that why you recoil from the heat and sun?
Do you feel trapped, contained, restrained?
Is that why you weep and seep
through window panes?
Does it thrill you, fulfill you,
to dissipate—evaporate?
Does it feel weird to leave the earth,
to rise above the rest?
Do you get dizzy hovering in the heavens?
Are you afraid to fail—to fall?
Aren't we all.
Aren't we all.


### AY AGUA, MI AMIGA
por **Guadalupe Garcia McCall**

¿Tienes miedo?
¿Es por eso que corres
hacia ríos, arroyos, lagos?
¿Estás más segura encharcada en estanques?
¿Te duele cuando hierves?
¿Es tu rabia que rueda y se agita?
¿Por eso retrocedes del calor y del sol?
¿Te sientes atrapada, contenida, refrenada?
¿Es por eso que lloras y goteas
sobre el vidrio de las ventanas?
¿Te emociona, te realiza,
poder disiparte, evaporarte?
¿Te sientes rara al dejar la tierra,
al ascender más alto que los demás?
¿Te mareas flotando en los cielos?
¿Tienes miedo de fracasar, de caer?
Nosotros también.
Nosotros también.

### Take 5!

1. **Read this poem aloud while playing a video and audio soundtrack of water images and sounds in the background,** such as those found at Calm.com.

2. Since this is a bilingual poem, it's the perfect opportunity to **invite anyone who speaks Spanish (in your class, school, or community) to join you by reading the Spanish poem** before or after you read the poem in English.

3. **Invite students to find or draw an image for one of the questions asked in the poem.** Then arrange them sequentially next to a copy of the poem to show a visual interpretation of the poem.

4. **Use this poem to discuss the continuous movement of water above and on the surface of Earth through the water cycle.** Challenge students to cite specific examples from the poem. Resources like the U.S. Geological Survey provide helpful models for exploring the processes in the water cycle, including evaporation, condensation, and precipitation. See GA.water.USGS.gov/edu/watercycle.html.

5. Link this poem with another about water, **"Ocean Engine" by Leslie Bulion** (5th Grade, Week 16, page 244), and with selections from *How to Cross a Pond: Poems about Water* by Marilyn Singer.

*THE POETRY FRIDAY ANTHOLOGY FOR SCIENCE*

## Week 17: Weather & Climate

**Fourth Grade**

### Take 5!

1. **Project a meteorological map as a backdrop** for reading this poem aloud. Locate a map for your own area at Weather.com.

2. Share the poem again, and **invite students to chime in on the important final line** as you read the rest of the poem aloud.

3. **Work with students to identify "weather words" in this poem** (e.g., *sun, sleet, weather patterns, meteorologist, rain, sunny, hot, humid, cold, wet, dry, weathercaster, cold front*). Discuss their meanings and decide which apply to today's weather in particular.

4. **Challenge students to measure and record recent changes in weather** and make predictions using weather maps, weather symbols, and a map key. Use online resources like Weather.com as well as local news sites.

5. Connect this poem with another about weather news, **"This Week's Weather" by Janet Wong** (3rd Grade, Week 17, page 165), and selections from *Seed Sower, Hat Thrower: Poems about Weather* by Laura Purdie Salas or *Weather Report* by Jane Yolen.

## Weather Map
### by Joan Bransfield Graham

A crazy quilt of suns and sleet,
weather patterns that repeat,

the meteorologist paints away—
a chance of rain, a sunny day.

Hot and humid, cold, wet, dry—
the weathercaster's practiced eye

checks for clues, consults the chart,
knows the online sites by heart,

plots the future from the past.
How long will that Cold Front last?

Arrows show its bold advance
in the daily Weather Dance.

Clouds and suns are rearranging . . .
weather's face is always changing.

## Week 18: Forces of Nature

**Fourth Grade**

### Hurricane Hideout
by **Janet Wong**

The weather reporter
says it might be
a really bad hurricane,
Category 3.
Her words
are jumbling up
in a ball:
storm surge,
flood watch,
winds,
eye wall,
low pressure,
cyclone,
large heat engine.

All I know is
I'm going to get
IN
the safest part
of our house,
the tub.
I'm going in there
with the stuff I love:
radio, pillows,
candy, light, book.
Watch the news
but don't make me look.
What? You filled the tub
to the top
in case our water
gets shut off?

No problem.
I can deal with it.
You'll find me
in the hall closet!

### Take 5!

1. **After reading this poem aloud, share a video that shows how hurricanes form along with hurricane footage.** National Geographic offers a helpful example, "Hurricanes 101," found at Video.NationalGeographic.com/video/kids/forces-of-nature-kids/hurricanes-101-kids/.

2. Share the poem again, and **invite students to say the list of hurricane words** (*storm surge, flood watch, winds, eye wall, low pressure, cyclone, large heat engine*) **as well as the list of *stuff I love* words** (*radio, pillows, candy, light, book*) while you read the rest of the poem aloud.

3. **Time to discuss emergency preparedness,** whether you live in a hurricane-prone area or not. Review emergency procedures for school or home together.

4. Talk with students about how the surface of the earth is constantly changing. **Research how wind and rain combine to create tropical depressions and storm surges** that can result in hurricanes, cyclones, and typhoons. One excellent resource can be found at Weather.com/outlook/weather-news/hurricanes/articles/tropical-storm_2010-08-04.

5. Connect this poem with another about violent storms, **"Tornado!" by Carole Gerber** (3rd Grade, Week 18, page 166), and with the informative nonfiction books *Hurricanes!* by Gail Gibbons or *Hurricanes* by Seymour Simon.

*The Poetry Friday Anthology for Science*

## Week 19: Soil & Land

**Fourth Grade**

### Take 5!

1. If possible, **bring a carrot or two or even a handful of dirt or soil as your poetry prop** to show after reading this poem aloud.

2. Share the poem again, and this time **invite students to say the words that Grampa says**—"*We all got to eat / a ton of dirt*—while you read the rest aloud.

3. For discussion: ***How does the soil feed us and how do we feed it?***

4. **Use this poem to talk about the food chain within a garden** and how all living organisms have basic needs that must be met for them to survive within their environment. We nurture the soil (with sensible planting, fertilization, crop rotation) and the soil provides food for us, etc.

5. Link this poem with another about the cycle of sun, soil, and growth, **"Sun-Kissed / Besado por el sol" by Guadalupe Garcia McCall** (3rd Grade, Week 23, page 171), and look for selections from *Busy in the Garden* by George Shannon.

## What We Eat
### by Joseph Bruchac

It has been said
we are what we eat.

Perhaps that is so
but what I think now
is that in the end
it all comes down
to the simple words
my Indian grandfather
spoke one late summer
as we knelt
in the garden
to pull up
sun-gold carrots.

Brushing them
on his sleeve,
Grampa said to me
"We all got to eat
a ton of dirt"
his words taking me
back to the soil
that feeds us,
that we feed.

## Week 20: Natural Resources

### Solar Power
by **Susan Blackaby**

Solar power! Feel the heat!
Light the lights along the street,
run the engines, fuel the cars,
turn the turbines with a star!

Quick! Let's build a head of steam,
fire up some clean machines,
set in motion cranks and cams,
swirling gears and pumping dams.

Solar cells turn light to juice—
electron transfer on the loose!
Tap this energy in space!
The sun can win Earth's resource race.

### Take 5!

1. If a window is nearby, **stand by the window while reading this poem aloud,** highlighting the importance of the sun and solar power in this poem.

2. Share this poem again, and **invite students to say the first line of each stanza** while you read the rest of the poem aloud.

3. **Survey students on which** solar powered objects they have experienced collectively (house, car, light, toy, calculator, and so on). Make a simple chart documenting the results.

4. **This poem can jumpstart a discussion of Earth's renewable resources**, including air, plants, water, and animals; nonrenewable resources such as coal, oil, and natural gas; and the importance of conservation. Work together to find more information on using the sun as a source of electricity at EIA.gov/kids/energy.cfm?page=solar_home-basics-k.cfm.

5. Match this poem with another about solar power, **"Auntie V's Hybrid Car" by Janet Wong** (Kindergarten, Week 20, page 48), and the nonfiction photo-essay book *The Sun* by Seymour Simon.

## WEEK 21: ECOSYSTEMS

**FOURTH GRADE**

### Take 5!

1. Before sharing this poem, take a moment to **encourage students to close their eyes and imagine a tall, swaying tree, among many tall trees, surrounded by many branches, plants, puddles, and insects.** Then continue by reading this poem aloud.

2. Read the poem again and **encourage students to join in on the repeated line *No space is ever wasted*** while you read the rest of the poem aloud.

3. Challenge students to work in pairs or trios to **research images of the plants or animals mentioned in the poem** (*tree, air plant, tadpole, insect, crab*), and then post all the pictures in order corresponding to the poem.

4. Use this poem to talk about how **living organisms within an ecosystem interact with one another and with their environment.** Consider the interdependent relationships described in the poem between the tree, air, plant, tadpole, insect, and crab. Consult the student-created website at BluePlanetbiomes.org/rainforest.htm.

5. Revisit the poem **"A Biological Community / Una comunidad biólogica"** also by **Margarita Engle** (4th Grade, Week 4, page 192), and look for additional selections in *Chatter, Sing, Roar, Buzz: Poems about the Rain Forest* by Laura Purdie Salas.

### TROPICAL RAIN FOREST SKY PONDS
by **Margarita Engle**

No space is ever wasted.
Each species must find
its own niche.

At the top of a swaying tree, air plants cling
to high branches, seeking sunlight.
Their dangling roots absorb moisture
from drifting mist, instead of soil.
At the base of each leaf of an air plant,
inside a small puddle of rain,
tadpoles turn into frogs,
insects swim,
and little crabs clack
tiny claws.

No space is ever wasted
in this forest
of surprises!

## WEEK 22: ADAPTATIONS & TRAITS

**FOURTH GRADE**

### GRAFTING
by **Janet Wong**

We have an old red apple tree
here in the yard of this new house
but the apples don't taste tart enough.
(We like them tart for pie.)
Dad says it's an easy fix.
Today we drove to our old house
and asked to cut some branches from
the apple tree we always loved.
They look like plain bare sticks to me
but Dad says they're just what we need.
We'll graft them onto our tree here,
cut and line the pieces up,
tie them tight and seal with wax,
keep it moist, and hope and hope.
If the grafts grow fine, come harvest time,
we'll have our tart apple pie!

### Take 5!

1. Add a bit of fun to sharing this poem with a poetry prop—**show a branch or an apple before reading the poem aloud.**

2. Share the poem again, and **invite students to join in on the two phrases attributed to Dad**—*it's an easy fix; they're just what we need*—while you read the rest of the poem aloud.

3. For discussion: *If you could graft two fruits together, which would you choose and why?*

4. **This is an example of selecting for genetic traits,** which improves the characteristics exhibited by the apples that are combined. Adaptations take place over thousands of years as the fittest within a species live to reproduce. Students can look up grafting techniques at sites like SaltSpringAppleCompany.com/Grafting-Apple-Trees.htm.

5. For another poem about growing plants, seek out **"First Science Project"** by **Lesléa Newman** (1st Grade, Week 26, page 94), and look for more tree poems in *Winter Trees* by Carole Gerber.

## WEEK 23: CYCLES

### FOURTH GRADE

## WINDFALL IN THE ANDREWS FOREST
by **Joseph Bruchac**

The way the giant Douglas fir
leaned after five centuries
showed the way
wind wanted it to go.

Wide roots, spread
into the soil like hands
were not enough to hold.

It crashed down through the canopy,
scattered branches over the stream,
needles and Old Man's Beard lichen
fluttering down like green rain.

The small trees below,
yews, hemlocks, and alders,
were not net enough to slow it.

But the earth and its stones were stronger,
for when the tree struck,
its great trunk broke,
its bark was shed like an overcoat,
and its layers of growth split to splinters.

### Take 5!

1. After reading this poem aloud slowly and quietly, **locate the Andrews Forest on a map**. See Andrewsforest.OregonState.edu.

2. Share the poem again, and **invite students to join in on reading the climactic middle stanza together** (*It crashed down through the canopy...fluttering down like green rain*). You may need to explain that *Old Man's Beard* is a kind of lichen that grows from tree branches.

3. **Talk about the different kinds of trees mentioned in the poem** (*Douglas fir, yew, hemlock, alder*) and use books or sites like Arborday.org to learn about tree identification and trees in your area.

4. Use this poem to **talk about how all living organisms have a life cycle, including trees.** Challenge students to identify stages of the tree's life cycle described in the poem. An excellent PowerPoint slideshow is available at TinyURL.com/lgqxw76.

5. Link this poem with another tree poem, **"Rings Not Letters" by Juanita Havill** (2nd Grade, Week 24, page 132), and with selections from *Poetrees* by Douglas Florian.

## WEEK 24: PATTERNS
### FOURTH GRADE

### CICADA MAGIC
by **Heidi Mordhorst**

First, the news:
    cicadas are coming!

Then, the holes:
    one, three,
    fivethirteenseventeenfifty-oneseventy-three
    hundreds of little eruptions in the mud.

Next, the shells:
    brittle, hollow, yellow-brown;
    perfect casts, fresh fossils:
    blessedly motionless.

Not for long:
    the first ones appear
    with Martian-red eyes
    at face-height on the front porch.
    They litter the pavement,
    scatter the windshield.
    The lawn crunches underfoot.

And yet:
    they are friendly,
    allowing us to catch, keep,
    compare, even wear them.

And now, the song:
    subtle thrumming at morning,
    slowly swelling to a throb that meets
    the beating of the sun all afternoon.

### Take 5!

1. **Play cicada sounds in the background as you read this poem aloud.** One source is FreeSound.org.

2. Follow up with another reading of the poem, and **invite students to read the first line of each stanza** while you read the rest of the poem aloud.

3. Work together to **research what purposes cicadas might serve** in the environment.

4. **Use this poem to explore and illustrate the life cycle of the cicada.** What factual details does the poet provide? Find additional information at CicadaMania.com.

5. Pair this poem with **"Cicada / Chicharra" by Guadalupe Garcia McCall"** (5th Grade, Week 23, page 251), and look for more poems by Heidi Mordhorst in *Pumpkin Butterfly: Poems from the Other Side of Nature.*

## Week 25: Human Body

**Fourth Grade**

### Take 5!

1. This poem lends itself to pantomime while reading, if you're ready for hopping, push-ups, and more. **Or simply run in place while reading this poem aloud, and then pretend to take your pulse at the end.**

2. Present the poem again, and **invite students to say all the number words** (*50, half, 90, 80, one hundred sixty-two*) while you read the rest of the poem aloud.

3. **Work together to research normal pulse rates at various ages and during different activities.** One resource is My.ClevelandClinic.org/heart/prevention/exercise/pulsethr.aspx.

4. Talk about how scientists use a variety of tools to collect information and conduct investigations. In studying the human body, one measure of overall health is pulse rate. **Demonstrate how to "take a pulse" and compare results while resting or exercising.**

5. For another poem about measuring reactions, look for **"Zapped!" by April Halprin Wayland** (3rd Grade, Week 7, page 155), and seek out poems by Jack Prelutsky in *Good Sports: Rhymes about Running, Jumping, Throwing, and More*.

## Moving for Five Minutes Straight
### by **Betsy Franco**

For 50 secs,
we hop around,
then switch
for 50 more.

For half a minute
we all lie down
for push ups
on the floor.

For 90 secs,
we get in rows
to do our
jumping jacks.

For 80 secs,
we're on our feet,
all sprinting
up and back.

But when we
hear the buzzer sound,
we freeze.
We're glad to stop.

My pulse:
one hundred sixty-two!
I think my heart will
pop!

## WEEK 26: KITCHEN SCIENCE

### FOOD FOR THOUGHT
by **Robyn Hood Black**

You won't find a character, setting, or plot
on the side of the cereal box Dad bought.

But wait! There's still something tasty to read.
The **food label** has information you need.

*Ingredients* tell you what is inside.
(See sugar and salt? They were trying to hide.)

Your body needs **protein, carbohydrates,** and **fat**.
A good bit of this, just a little of that.

**Vitamins** help keep you active and strong—
**minerals**, too, when they tag along.

Check out the **calories** per **serving size**.
Then make a choice that is healthy and wise!

### Take 5!

1. If possible, add some fun to sharing this poem with a poetry prop—**show a box of cereal or the ingredients panel from any food package** before reading the poem aloud.

2. For a follow-up reading, **invite students to say the key words in bold** while you read the rest of the poem aloud.

3. **Share a video on interpreting nutrition facts food labels,** found at KidsHealth.org/kid/stay_healthy/food/labels.html. Talk about how kids can make healthy choices based on this information.

4. **Use this poem to discuss how consumers can evaluate product claims** found in advertisements and labels for food. Compare the labels on various products to assess their nutritional value, and then place them into food group sections on MyPlate using this site: MyPlate.gov/food-groups/.

5. Link this with another poem about measuring food ingredients, **"Breakfast Alchemy" by Mary Quattlebaum** (3rd Grade, Week 26, page 174), and look for "recipe" poem books such as *Guacamole: Un poema para cocinar/ A Cooking Poem* by Jorge Argueta.

## Week 27: Video Technology

**Fourth Grade**

### Virtual Adventure
by **Renée M. LaTulippe**

Yesterday I scaled some peaks.
Looky here: wind-chapped cheeks!

Right after lunch, I rode a gnu,
caught cuckoo birds in Katmandu.

Snowboard? Check. Windsurf, scuba.
After dinner? Played a tuba.

Safari in the Serengeti,
tango with a sweaty yeti.

I can do most anything—

from biking in downtown Beijing
to wrestling deep-sea squid-eos—

with my green screen videos.

### Take 5!

1. After reading this poem aloud, **use Voki.com to create a simple avatar for yourself** and encourage the students to create their own avatars, too.

2. Read the poem aloud again, and **invite students to say the crucial last line together** while you read the rest of the poem.

3. **Share a video of a student-made green screen poetry project** available at PoetryforChildren.blogspot.com/2013/05/ms-neelands-green-screen-mo-po-poetry.html.

4. Talk about how digital resources enable us to have virtual experiences without leaving our desks, and how we can use those same tools to create our own virtual worlds. **Use this poem to identify and discuss each place featured and talk about the role of "green screen" technology in presenting places virtually.**

5. Compare this poem with another about traveling via technology, **"Hello, Hello!" by Janet Wong** (Kindergarten, Week 27, page 55), and read about actual explorers in *Trailblazers: Poems of Exploration* by Bobbi Katz.

## WEEK 28: MACHINES

### MICROWAVE OVEN
by **Janet Wong**

It takes twenty minutes
to make a batch
of my favorite
blueberry pancakes
from scratch.

On weekdays we need
every minute of sleep
so we microwave
pancakes:
one minute—beep!

But watch out:
the pancakes get HOT
in the middle—
much hotter than when
they come off the griddle.

Mom thinks
microwaves cook
from the inside-out
but that's not true:
I've learned about

how microwaves
get into the food
just a bit and activate
water molecules.
Once hit,

they vibrate
and bounce
to make lots of heat.
Electromagnetic radiation—sweet!

### Take 5!

1. Before sharing this poem, **survey students on which** cooking modes they have experienced (microwave, stovetop, oven, campfire). Then read the poem aloud.

2. Share the poem again, and **invite students to chime in on the exclamation words *beep* and *sweet*** while you read the rest of the poem aloud.

3. Talk about the importance of **home and school safety procedures regarding the use of microwaves, especially in the handling of hot foods.** Search HealthyChildren.org for "microwave safety."

4. Use this poem to **talk about different forms of energy**, including mechanical, sound, electrical, light, and heat/thermal. Share NASA resources for understanding electromagnetic energy found at MissionScience.NASA.gov/ems/01_intro.html.

5. Revisit the poem **"Food for Thought" by Robyn Hood Black** (4th Grade, Week 26, page 214) or compare with another poem featuring microwave cooking, **"Five O'Clock Rush" by F. Isabel Campoy** (3rd Grade, Week 28, page 176). Share microwave recipes in *Cook It in a Cup* by Julia Myall and Greg Lowe.

## WEEK 29: BUILDING THINGS

**FOURTH GRADE**

### Take 5!

1. **To set the stage for this poem, stack several blocks** (or books or other stable materials) to suggest a small building. Then read this poem aloud, altering your voice slightly for the dialogue spoken by Grandpa and by the poem narrator.

2. Share the poem again, and **pause for students to chime in on the crucial line** *once you rush it, you crush it.*

3. For discussion: *When is it most important to read the directions?*

4. **Use this poem to highlight the steps involved in problem solving and decision making.** Identify examples regarding the problem and explain the steps toward the solution, including sorting materials, reading directions, establishing a base, proceeding methodically, and even maintaining calm.

5. Revisit a previous poem about building **"Computer Models" by Janet Wong** (4th Grade, Week 8, page 196), and check out *Monumental Verses* by J. Patrick Lewis; *Castles: Old Stone Poems* by Rebecca Kai Dotlich and J. Patrick Lewis; or *Sacred Places* by Jane Yolen. For an interesting twist, look for *Unbuilding* by David Macaulay.

### FOUNDATION (DON'T RUSH IT!)
by **Charles Waters**

Where to begin?
These pieces are all chunked together.
Directions read
like a pile of hieroglyphics.
I cross my arms and stew.
Grandpa walks by.
"What's wrong?" he asks.
"I don't know where to begin," I reply.
"A good rule of thumb
is to start at the bottom,
have a good base.
That's called a foundation."
He reads the directions slowly as
we begin to slot pieces together
on the wooden floor.
"Being patient is key—remember,
once you rush it, you crush it."
This building starts rising like helium,
turning into the skyscraper of my dreams.

# Week 30: Science Fair

### Fourth Grade

## Science Fair
by **Irene Latham**

The graphics
I created and pinned
to the felt board

explain why my eyes
could never be brown,
my hair only blond.

I wonder if Mendel's
theory of genetics
also applies to why

I'm shy
and can speak
to the judges

only in a quavery voice
that betrays my shaky
hands and knees.

### Take 5!

1. Add a personal dimension to this poem by **pantomiming the actions suggested by the poem while you read it aloud** (point to your eyes and hair, shrug your shoulders, act shy, show shaky hands and knees).

2. Share the poem again, and **invite students to join you in reading the pivotal middle stanza of the poem** (*I wonder if Mendel's / theory of genetics / also applies to why*) while you read the rest of the poem aloud.

3. For discussion: *Which traits do we inherit from our family's biology (eye color, hair color) and which from growing up around family members (shyness, courage)?*

4. Use this poem to **talk about how some likenesses between parents and offspring are inherited,** or passed from generation to generation, such as eye color in humans or shapes of leaves in plants. Other likenesses are learned, such as table manners or hobbies. Work together to learn more about Gregor Mendel and his work. One resource is FamousScientists.org.

5. Match this poem with another about genetic traits, **"Inherit Tense" by Charles Ghigna** (Kindergarten, Week 24, page 52), and look for *Am I Naturally This Crazy?* by Sara Holbrook or *Gregor Mendel: The Friar Who Grew Peas* by Cheryl Bardoe.

## WEEK 31: FAMOUS SCIENTISTS

### ALBERT EINSTEIN
by **Julie Larios**

Einstein's mind
matched his hair.
Both had something
energetic blow through them
at the speed of light squared.
In fact Einstein's massive frizz—
brain-wise, hair-wise—
was absolutely atomic.

### Take 5!

1. **To set the stage for this poem, show a close-up photo of Albert Einstein.** One source is the photo gallery at Einstein.biz. Then read the poem aloud slowly and dramatically.

2. Share the poem again **and invite students to join in on the last two words of the poem (*absolutely atomic*).**

3. **Talk with students about the pun and parallel** the poet makes between Einstein's brilliant scientific mind and the description of his famously wild or "explosive" hair.

4. **Work together to research Einstein's many contributions to science,** along with the history of science and possible science careers. One helpful resource is Biography.com.

5. Make another connection with the topic of atoms with **"Imagine Small" by Eileen Spinelli** (2nd Grade, Week 9, page 117) or **"Think of an Atom" by Buffy Silverman** (5th Grade, Week 9, page 237). For more poems about people who have made a difference, look for *Dare to Dream...Change the World* edited by Jill Corcoran or seek out the biographical picture book *On a Beam of Light: A Story of Albert Einstein* by Jennifer Berne.

## Week 32: More Famous Scientists

**Fourth Grade**

### Hawking Time
by **James Carter**

Stephen Hawking
hear my rhyme.
Tell me what
you think of time.

Is it straight?
Does it bend?
When it's over,
well, what then?

Was there time
when time was not?
Who winds up
the cosmic clock?

Mr. Hawking,
if you please.
Do solve all
these mysteries.

I know what you
might say to me.
"Go down to
the library."

### Take 5!

1. **Set the stage by sharing a bit of information about Stephen Hawking.** Check out Hawking.org.uk. Then read the poem aloud, pausing at the end of each question.

2. Next, read the poem aloud again, and **invite students to join in on their favorite question line or couplet** while you read the rest of the poem aloud.

3. Seek out the fun science-based novels by Stephen Hawking and his daughter Lucy Hawking and **share a chapter or excerpt** (e.g., *George's Secret Key to the Universe, George's Cosmic Treasure Hunt, George and the Big Bang*). Audio clip excerpts are available at Amazon or Books.SimonandSchuster.com.

4. **Consider Hawking's place in the pantheon of recent scientists.** Consult TimelineIndex.com/content/select/820/1023,820?pageNum_rsSite=20&totalRows_rsSite=321 and talk about the contributions of various scientists within the last 100 years.

5. For another poem on the topic of time, share **"My Project for the Science Fair" by Kenn Nesbitt** (1st Grade, Week 30, page 98), and look for selections from *The Tree that Time Built: A Celebration of Nature, Science, and Imagination* edited by Mary Ann Hoberman and Linda Wilson.

*The Poetry Friday Anthology for Science*

## Week 33: Computers

**Fourth Grade**

### Take 5!

1. **For a poetry prop, show three photographs—your own or from online sources** like Google Images or iStock. Then read the poem aloud faster and faster, slowing down at the end.

2. Read this poem aloud again, and **invite students to join you on the important last two lines** *(I'm doing it all / AGAIN)*. Read those lines together slowly and with exaggerated emotion.

3. **Science projects can include technological investigations** as well as more traditional activities. Brainstorm more possibilities students might consider for science fair projects.

4. **Recording data and taking notes are an important part of scientific inquiry.** Use this poem to talk with students about at what points the "speaker" in the poem *should* have stopped to take notes during this photo experiment.

5. Revisit two previous poems about conducting experiments with **"Testing My Hypothesis" by Leslie Bulion** (3rd Grade, Week 5, page 153) and **"Meet Mr. Wizard" by George Ella Lyon** (3rd Grade, Week 6, page 154). Or seek out *Art Lab for Kids: 52 Creative Adventures in Drawing, Painting, Printmaking, Paper, and Mixed Media for Budding Artists of All Ages* by Susan Schwake.

## My Photo Experiment
### by Janet Wong

For my science fair project
I took three favorite photos.
I copied each file ten times.
Then I played with the settings,
adjusting saturation,
then scribbling down
notes of what I'd done.
I moved onto contrast,
exposure, and tint.
I got so excited
I just had to print.
I stopped taking notes
and I worked the slider bars
and I printed and I printed.
I got pretty far on my project—
I almost got to the end—
when I realized . . .
without notes,
I'm doing it all
AGAIN.

## Week 34: Science Careers

### Rocky Rescue
by **Robyn Hood Black**

In the South Pacific,
Lord Howe Island has a tale
of how a giant stick bug,
thought extinct, might prevail.

"Land lobsters" as they're called
had lots of woe in store
when, back in 1918,
a ship wrecked on their shore.

Rats skittered from the boat
and found the black bugs tasty.
"They're gone!" the experts said. "Each one!"
—a conclusion that proved hasty.

For not so long ago,
some scientists, at night,
climbed a sea stack miles away
and found an awesome sight.

*Look! The giant stick bugs!*
They counted twenty-four.
Now with help from science,
there are many, many more.

### Take 5!

1. Read this poem aloud with a brief pause between each stanza of the poem story. **Follow up with this one-minute video news story** that highlights the facts and shows the stick bug close up: YouTube.com/watch?v=rcvRb_3dp8c.

2. Read the poem together, **inviting students to read the portions within quotation marks**—*"They're gone!" "Each one!"* and *Look! The giant stick bugs!*—while you read the rest of the poem aloud.

3. **Work together to find Lord Howe Island (near Australia) on the map.** For images, see LordHoweIsland.info.

4. Use this poem to talk about how scientists help save animals from extinction. **Explore the science museum resources in your area for learning more about science and science careers.** Go to CareerCornerstone.org/muscenters.htm.

5. Link this poem with another about extinction, **"The Lament of Lonesome George" by Jane Yolen** (2nd Grade, Week 35, page 143). Also look for more poems about insects with *Hey There, Stink Bug!* by Leslie Bulion; *Bugs: Poems about Creeping Things* by David L. Harrison; and *Insectlopedia* by Douglas Florian.

## Week 35: Future Challenges

**Fourth Grade**

### Take 5!

1. **Bring some chocolate as your poetry prop** to set the stage for reading this poem aloud.

2. Share the poem again, and **encourage students to chime in on the three chocolate words—*cacao, cocoa, chocolate*—as they pop up in the poem while you read the rest aloud.

3. **Work together to research the process of growing cacao and turning it into chocolate.** Cadbury offers a helpful video at YouTube.com/watch?v=nfczfl0G_30.

4. Use this poem to talk about how all living organisms within an ecosystem interact with one another and with their environment. **Identify each of the organisms included in the rain forest ecosystem in this poem** (trees, cacao plants, toucans, parrots, monkeys, marmosets, orchids, butterflies). How do changes in the ecosystem (such as deforestation) affect the food web?

5. Revisit another poem about the rain forest, **"Tropical Rain Forest Sky Ponds"** also by Margarita Engle (4th Grade, Week 21, page 209) and look for more food poems in *Yum! Mmmm! Que Rico!: America's Sproutings* by Pat Mora or *Lettuce Introduce You: Poems about Food* by Laura Purdie Salas.

## SHADE-GROWN
### by Margarita Engle

One village away: a deforestation disaster.

But here there is no slash-and-burn.
No clear-cuts.
Beneath towering rain forest trees,
farmers plant cacao.

Toucans and parrots weave nests.
Monkeys and marmosets play.
Orchids bloom.
Butterflies flutter.

A plentiful harvest of cocoa pods
will be roasted to make sweet chocolate.
An eco-agricultural success story.
Delicious!

## WEEK 36: FUTURE DREAMS

### FOURTH GRADE

### THE REAL THING
by **Linda Sue Park**

They say we're getting closer
to a real invisibility cloak.
It's not ready yet,
but it will be, someday soon.

Would you want one?
Would you try it out at the store,
beg for one for your birthday,
put it on your wish list?

The first ones for sale
will be really expensive.
But they'll come down in price
a bit at a time

until maybe you could save
(allowance, babysitting, lawn mowing)
for the basic model—the IC 100,
which doesn't quite cover your shoes.

It wouldn't take long
before lots of people had them.
There would be rules
about not bringing them to school.

New sports leagues!
IC basketball. IC soccer.
It sounds like "I See,"
but of course, you wouldn't,

except for odd ripples:
the ghost of a thumb,
a flicker of cleats,
an eyebrow floating in midair.

### Take 5!

1. Read this poem aloud pausing briefly between stanzas. **Then show a brief example video of real experiments with invisibility cloaks.** One sample is at YouTube.com/watch?v=PD83dqSfC0Y.

2. Share the poem again, and **invite students to chime in on the crucial phrase *the IC 100* while you read the rest of the poem aloud.**

3. For discussion: *What kinds of rules would we need if everyone had invisibility cloaks?*

4. This poem encourages us to look to the future and imagine what kinds of inventions scientists will create next. **Work together to brainstorm and forecast possibilities and talk about what steps might be needed for the creation of an innovative process or product.**

5. Pair this poem with another fun future-forecasting poem, **"Invention Intentions" by Kristy Dempsey** (3rd Grade, Week 34, page 182).

# Poems for Fifth Grade

# NGSS Science and Engineering Practices: Fifth Grade

*These practices form the foundation of disciplinary literacy in science and integrate reading, writing, listening, and speaking skills from the language arts. Here we indicate which weekly poems emphasize which science and engineering practices at each grade level.*

| PRACTICE | POEM |
|---|---|
| Asking questions and defining problems | Weeks 3, 27, 36 |
| Developing and using models | Weeks 9, 13, 15, 24 |
| Planning and carrying out investigations | Weeks 1, 6, 8 |
| Analyzing and interpreting data | Weeks 4, 11, 12, 25 |
| Using mathematics and computational thinking | Weeks 7, 26 |
| Constructing explanations and designing solutions | Weeks 5, 16, 17, 20, 23, 28, 29, 31, 33 |
| Engaging in argument from evidence | Weeks 19, 30, 34 |
| Obtaining, evaluating, and communicating information | Weeks 2, 10, 14, 18, 21, 22, 32, 35 |

# Fifth Grade

| | | |
|---|---|---|
| week 1 | Scientific Practices | Scientific Inquiry *by Susan Blackaby* |
| week 2 | Lab Safety | Welcome to the Science Lab *by Heidi Bee Roemer* |
| week 3 | Ask and Ask Again | Paper Airplanes *by Janet Wong* |
| week 4 | Observations | Playground Physics *by Jeannine Atkins* |
| week 5 | Predictions & Hypotheses | Accidentally On Purpose *by Linda Sue Park* |
| week 6 | Investigations | Designing an Experiment *by Avis Harley* |
| week 7 | Data | Going Bananas *by Heidi Bee Roemer* |
| week 8 | Tools of Science | That Dish Thing *by Virginia Euwer Wolff* |
| week 9 | Matter | Think of an Atom *by Buffy Silverman* |
| week 10 | More Matter | Elemental *by Jeannine Atkins* |
| week 11 | Force, Motion & Energy | No Penguins Here *by Michael Salinger* |
| week 12 | More FM&E | At the Speed of Light *by Shirley Smith Duke* |
| week 13 | Light & Sound | Teacher's Look *by Shirley Smith Duke* |
| week 14 | Space | Comet Hunter *by Holly Thompson* |
| week 15 | Sun, Earth & Moon | The Shadow Grows *by Laura Purdie Salas* |
| week 16 | The Water Cycle | Ocean Engine *by Leslie Bulion* |
| week 17 | Weather & Climate | Climate Versus Weather *by Joan Bransfield Graham* |
| week 18 | Forces of Nature | Harbor Wave at Hilo *by Carole Boston Weatherford* |
| week 19 | Soil & Land | Soil Inventory *by Kate Coombs* |
| week 20 | Natural Resources | Resources Rule! *by Susan Blackaby* |
| week 21 | Ecosystems | Cool Food for Thought *by Sara Holbrook* |
| week 22 | Adaptations & Traits | What Is a Foot? *by Jane Yolen* |
| week 23 | Cycles | Cicada/Chicharra *by Guadalupe Garcia McCall* |
| week 24 | Patterns | Patterns in Nature *by Shirley Smith Duke* |
| week 25 | Human Body | Let Me Join You *by Heidi Bee Roemer* |
| week 26 | Kitchen Science | Thirsty Measures *by Heidi Bee Roemer* |
| week 27 | Video Technology | Frames Per Second (fps) *by Janet Wong* |
| week 28 | Machines | Soda Machine Bite *by Jacqueline Jules* |
| week 29 | Building Things | The Great Pyramid of Giza *by Laura Purdie Salas* |
| week 30 | Science Fair | Tell It to the Court *by Janet Wong* |
| week 31 | Famous Scientists | Shen Kuo *by Janet Wong* |
| week 32 | More Famous Scientists | I Will Be a Chemist/Voy a ser químico *by Alma Flor Ada* |
| week 33 | Computers | Printing, Pressed Beyond Words... *by Robyn Hood Black* |
| week 34 | Science Careers | A New Dinosaur *by Marilyn Nelson* |
| week 35 | Future Challenges | Titan in Man's Seaweed *by Michael J. Rosen* |
| week 36 | Future Dreams | My WristRobot Pack *by Carmen Tafolla* |

THE POETRY FRIDAY ANTHOLOGY FOR SCIENCE

"To **pay attention**, this is our endless and proper work."

❦ Mary Oliver ❧

## Week 1: Scientific Practices

**Fifth Grade**

### Take 5!

1. Before sharing this poem, **alert students to listen for four particular science terms that are essential to scientific exploration** (*hypothesis, observations, data, results*). Read the poem aloud and then discuss those terms together.

2. Share the poem again, **and invite students to say those key science words in bold** while you read the rest of the poem aloud.

3. Use this poem to talk about each step in the inquiry process. **Collaborate to create a quick glog, a digital interactive poster (using Glogster.com), pulling together images to illustrate these key words.** Show the students the choices of text, fonts, color, graphics, and even animation, if possible, while you input those items and create the finished product.

4. **Point out that all scientists use regular scientific methods** during laboratory and outdoor investigations. They analyze, evaluate, and critique scientific explanations by using empirical evidence, logical reasoning, and experimental and observational testing, including examining all sides of the evidence.

5. Pair this poem with **"Scientific Steps" by Cynthia Cotten** (A Poem for Everyone, page 23) and look for *Spectacular Science: A Book of Poems* edited by Lee Bennett Hopkins.

### Scientific Inquiry
by **Susan Blackaby**

Scientists are like explorers,
using what they know and see
to blaze a trail that, step by step,
will lead to new discoveries:

Formulate, distill, and focus,
narrow down, define the gist,
determine scope and pinpoint locus—
this is your **hypothesis**.

Gather all the stuff you need,
to put in play the machinations.
Document the happenings—
these comprise your **observations**.

What things change and what things stick?
Record the outcomes and effects.
Don't presume and don't predict—
collect the **data**: just the facts.

Combine the concrete things you see
with what you know and trials you test.
Interpretation is the key—
**results** are where you end the quest.

## WEEK 2: LAB SAFETY

**FIFTH GRADE**

### WELCOME TO THE SCIENCE LAB
by **Heidi Bee Roemer**

Please quietly enter
the science lab center.
      Remember to follow the rules.

Lab work is awesome,
but always use caution
      with chemicals, burners, and tools.

You'll be given a pair
of goggles to wear,
      plus gloves, a lab coat or apron.

As you measure and pour
always be sure
      to wipe away spills at your station.

You may whip up a brew
of some gloppity goo;
      Never act on the foolish notion

of tasting the stuff—
you may go belly up!
      It might be a poisonous potion.

If you make a mistake
and a beaker should break,
      bring it straight to your teacher's attention.

But don't mess around
or you may be found
      on the receiving end of a detention.

What a success!
You've proven your test.
      Time to put your equipment away.

Wash your hands. Now you're through.
You've learned something new—
      and for lab safety you've earned an A!

### Take 5!

1. Add a bit of fun to sharing this poem with a poetry prop—**show an item of science lab equipment that is handy like goggles or beakers**—and then read the poem aloud.

2. Share the poem again, and **invite students to cheer with the final letter grade at the end of the poem** while you read the whole poem aloud.

3. Invite students to **choose one stanza to quick-sketch from this poem.** Create a quick collage of their sketches arranged around a copy of the poem.

4. **When it comes to science, it's important to begin by talking about science safety procedures.** Research together what your school, district, or community mandates for classroom and outdoor investigations. This might include wearing safety goggles, washing hands, using materials and equipment appropriately, etc. Then create a simple poster highlighting these guidelines.

5. Connect this poem with another about lab safety, **"Dinos in the Laboratory" by Kristy Dempsey** (4th Grade, Week 2, page 190).

# Week 3: Ask and Ask Again

## FIFTH GRADE

### Take 5!

1. Read this poem aloud with a pause between each stanza. **Then ball up a piece of paper at the end and toss it across the room!**

2. Share the poem again, and **invite students to join in on the last, exclamatory line,** *We won! We won!! We won!!!*

3. **Work together to investigate science competitions that students might consider participating in.** One resource is TryScience.org/parents/se_7.html.

4. Use this poem to **talk about how scientists often repeat investigations to increase the reliability of results,** and sometimes use creative thinking and innovative processes to solve a tricky problem. Challenge students to identify each step using examples from the poem.

5. Link this poem with another poem that poses many questions, **"Inquiry" by Cynthia Cotten** (3rd Grade, Week 3, page 151), or seek out more airplane poems in *Skywriting: Poems in Flight* by J. Patrick Lewis. For directions on making a variety of paper airplanes, consult the classic *Paper Airplane Book* by Seymour Simon.

## PAPER AIRPLANES
by **Janet Wong**

Each team in our class has twenty minutes
to make a paper plane that can fly the farthest.

One sheet of paper per plane.
No other stuff.
Five pieces of paper per team for models.
Each team works in a separate area. No spies.

We brainstorm.
Short and wide or long and thin?
Wing tips up or down or flat?
Pointed nose or squared off?

We make five models and test them all.
With one minute to choose our favorite,
our best plane flies straight into a wall
on its third test flight at the very same time
that our principal walks through the hallway
and steps on it. *Crunch!*
It is broken beyond repair.
Glenn crumples it into a ball and throws it.
It goes farther than anything else we made.
We have ten seconds left when—*ding dong!*—
a question pops into our minds.
A stupid question?
Maybe. But we run to ask our teacher anyway:
*Does it have to look like a regular plane?*

Kids laugh when they see our ball-plane.
But no one laughs when we jump and shout:
*We won! We won!! We won!!!*

## WEEK 4: OBSERVATIONS

### PLAYGROUND PHYSICS
by **Jeannine Atkins**

Boys on the seesaw learn about levers.
A girl on a swing pumps her legs, jumps, and lands.
Her sneakers slip and skid. She's grateful for gravity
and friction, which makes her sneakers stop in sand.

Mr. Newton's students play with a bat, a ball, force,
energy, and motion. The pitcher throws balls that spin.
One batter strikes out. The next sees the pitcher twist
his wrist. The batter swings, then sprints and grins.

Quieter scientists watch ants use their jaws
like levers, lifting long blades of grass.
Could ants make a seesaw? There's no time to ask.
Recess is over. The bell rings. Back to class!

### Take 5!

1. What is the perfect backdrop for this poem? A video of playground fun. **Play this video without sound effects while you read the poem aloud**: YouTube.com/watch?v=fpQTRqW-Juo.

2. Share the poem again, and **invite students to join in on the important last line** (*Recess is over. The bell rings. Back to class!*) while you read the rest of the poem aloud.

3. Use this poem to **create a list of possible playground activities** featuring two columns, one for the activity and another for the physics demonstrated in each activity (such as levers, gravity, friction, force, motion, and so on).

4. If possible, **encourage students to conduct their own observations of recess or playground activities** for one 30-minute period. Then gather to discuss their findings. Talk about how scientists analyze and interpret information to construct reasonable explanations from direct (observable) and indirect (inferred) evidence.

5. Connect this poem with **"Friction" by Sara Holbrook** (4th Grade, Week 12, page 200) and with selections from *Good Sports: Rhymes about Running, Jumping, Throwing, and More* by Jack Prelutsky.

## Week 5: Predictions & Hypotheses

**Fifth Grade**

### Take 5!

1. **As you read this poem aloud, include the head slap motion** both times the phrase *head slap* occurs. (Note that the phrase *head slap* is like stage directions and those words are not spoken out loud.)

2. Share the poem again, and **encourage students to do the *head slap* motion with you** as you read the rest of the poem aloud.

3. Conduct a bit of quick collaborative research to **learn more about crop rotation**. One resource is OrganicGardening.com.

4. Use this poem to talk about how scientists analyze data and interpret patterns to construct reasonable explanations from data that can be observed and measured. **Which crops are featured in the poem and what scientific benefits are cited?**

5. Match this poem with **"What We Eat" by Joseph Bruchac** (4th Grade, Week 19, page 207) and follow up with Jorge Argueta's "recipe" poetry book, *Sopa de frijoles/Bean Soup* or Pat Mora's food origin poems in *Yum! Mmmm! Que Rico!: America's Sproutings*.

## ACCIDENTALLY ON PURPOSE
by **Linda Sue Park**

A hundred thousand lifetimes ago,
somebody figured it out:

If you planted the corn in the middle,
and the beans around it, the bean vines
could climb up the cornstalk,

which meant less work because
you didn't need to put in poles
to support the beans, and hey,

it turns out that the corn takes nitrogen
out of the soil, but the beans put it back,
so the soil stays healthier.

Most of the time, we don't know
what we don't know. It's only later
that we realize it. *head slap*

They tasted good together, beans and corn.
Succotash. Tortillas with refried beans.
Baked beans and cornbread,

and surprise, surprise: each by itself
is missing something, but together
they make a complete protein.

*head slap* again.

The Poetry Friday Anthology for Science

# WEEK 6: INVESTIGATIONS

**FIFTH GRADE**

## DESIGNING AN EXPERIMENT: WILL A CAR ROLL FASTER DOWN A STEEPER SLANT?
by **Avis Harley**

Changing one letter at a time, you can turn *"slant"* into *"speed"*—

        SLANT
        SCANT
        SCENT
        SPENT
        SPEN**D**
        SPEED

Changing one variable at a time, you can see *"slant"* turn into *"speed"*—

    SCIENTIST AT WORK

      To test car speed on the   slant   of a slope,
too many variables leaves you   scant   hope!
        But you're on the right   scent   if you change one by one:
                        time well   spent   in experiment fun.
         Will the car roll faster?   Spend   a while on this quest.
                Each variable for   speed   needs its own test.

**Note:** This poem is a Doublet, a form based on a word game often appearing in newspapers and magazines for children: turning one word into another, one letter at a time. Avis Harley was probably the first to turn this game into a published poem, in her book *Fly with Poetry: An ABC of Poetry* (where she changed *sleep* into *dream*).

### Take 5!

1. As a backdrop for this poem, **play a video of a toy car speeding down a track** (without sound) while you read the poem aloud. One source is YouTube.com/watch?v=XhNm0K7_rxM.

2. Share the poem again, and **invite students to say the words all in capital letters with a pause between each** (*SLANT, SCANT, SCENT, SPENT, SPEND, SPEED* and then *SCIENTIST AT WORK*) while you read the rest of the poem aloud.

3. **Talk about the key words** (*slant, scant, scent, spent, spend, speed*) and the role of each in the science experiment described in the poem.

4. **Use this poem to talk about how to implement simple experimental investigations testing one variable,** in this case an investigation that tests the effect of force on an object. Identify the variable (*slant*) that is being tested in the experiment in this poem.

5. Pair this poem with **"Crazy Data Day" by Janet Wong** (2nd Grade, Week 7, page 115) and with selections from *Always Got My Feet: Poems about Transportation* by Laura Purdie Salas.

## Week 7: Data

*Fifth Grade*

### Take 5!

1. If possible, **bring a banana as a poetry prop** to show before reading this poem aloud.

2. Share the poem again, and **encourage students to say each number that occurs in the poem** (*five, 10, 18, 20, 42, 5, 20's*) while you read the rest of the poem aloud.

3. Work together to make a table reflecting the data provided in this poem. Feature five monkeys along the X axis, and bananas consumed along the Y axis.

4. Use this poem (and the table you created) to **discuss how scientists collect information by detailed observations and accurate measuring** and communicate valid conclusions in both written and verbal forms. Review the concepts of mean, mode, and median, too.

5. For another poem featuring data, share **"Moving for Five Minutes Straight" by Betsy Franco** (4th Grade, Week 25, page 213) and seek out *Count Me A Rhyme: Animal Poems by the Numbers* by Jane Yolen.

### Going Bananas
by **Heidi Bee Roemer**

Monkeys are competing on the beach beneath cabanas.
Going for the gold, five monkeys gobble up bananas.

Judges track the data, jotting down each monkey's score,
till the monkeys at the beach can eat not one banana more.

Judges post the totals of bananas each consumed:
10, 10, 18, 20, and a whopping 42!

Add up the bananas in this dandy data set.
Divide by 5 to find the **mean** and 20's what you get.

The **mode** refers to figures that appear once—and again.
The most repeated number in this data set is 10.

In math, position matters. Like a "monkey in the middle,"
the **median** is 18 in this silly data riddle.

*The Poetry Friday Anthology for Science*

## WEEK 8: TOOLS OF SCIENCE

### THAT DISH THING
by **Virginia Euwer Wolff**

Richard Julius Petri,
a stout and sturdy German,
was interested in illnesses
and microscopic vermin.
That vile tuberculosis,
its victims aching, moaning;
he didn't know his research
would someday be called cloning.

Distributing bacteria
in 1882,
he cultured, peered, and poured and stirred
to find out something new.
But lab equipment way back then
was quite unlike today:
no perfect vessel for the cells.
He pouted in dismay.

The tubes, the flasks, their angles
wouldn't let bacteria thrive.
Without a way to grow them,
how could the search survive?
Oh, woe for Richard Julius
but, acting on his wish,
he devised a flat container
and now we use the Petri dish.

### Take 5!

1. After reading this poem aloud, **show students what a petri dish looks like.** If you don't have an actual dish handy, use online images.

2. Share the poem again, and **invite students to chime in on the crucial last line** while you read the rest of the poem aloud.

3. **Expand on this poem story with the National Library of Medicine virtual exhibit on Civil War era medicine** at NLM.NIH.gov/exhibition/lifeandlimb/exhibition.html. Discuss the advancements that have been made in dealing with bacteria and infection.

4. Use this poem to **talk about how scientists use a variety of tools** like the petri dish to collect, record, and analyze information. Work together to learn more about Richard Julius Petri and his work. One resource is WhoNamedIt.com/doctor.cfm/1079.html, a dictionary of medical eponyms.

5. Follow up with another poem about treating disease, **"Shots! Shots! Shots!" by Joy Acey** (2nd Grade, Week 25, page 133). For more about inventors and innovators, look for *Eureka! Poems about Inventors* by Joyce Sidman, or for a poem about bacteria, seek out Sidman's *Ubiquitous*.

## WEEK 9: MATTER

### FIFTH GRADE

### THINK OF AN ATOM
#### by **Buffy Silverman**

Think of an atom
so tiny, so small—
a speck of the world
        a speck of us all,
a speck of the ocean
a speck of a fly
a speck of a mountain,
        a book or the sky.

Imagine that speck
growing wide, growing tall
        an atom as large as
your school or the mall.

The atom looks empty—
        almost nothing at all,
but there in the center
a tiny tight ball
of neutrons and protons
with mass and with weight.
How many of each?
        (for oxygen: eight!)

Its charges are balanced:
a proton adds one,
        —(each electron's a minus)
the neutrons add none.

Outside of the nucleus—
        that tight little ball—
the electrons are swirling
they're smaller than small
like pieces of dust
whizzing through space
a cloud of electrons
        in a zip-zapping race.

An atom is tiny—
astoundingly small—
Trillions lie here
on this dot that I scrawl.

---

### Take 5!

1. Before sharing this poem, alert students to listen for particular science vocabulary related to atoms. **Then read the poem aloud and make a list of all the words they can identify** (e.g., *atom, neutron, proton, mass, weight, oxygen, charge, electron, nucleus*).

2. Share the poem again, and **invite students to join in on the all-important final stanza.**

3. Challenge students to choose one stanza to illustrate from this poem. **Create a quick collage of their sketches or found images arranged around a copy of the poem.**

4. Use this poem to talk about how scientists classify matter based on physical and chemical properties. **Work together to find a model of an atom** from sources such as Chem4Kids.com/files/atom_structure.html.

5. Link this poem with others that explore various states of matter, such as **"Questions That Matter" by Heidi Bee Roemer** (3rd Grade, Week 9, page 157) and **"Pumpkin Experiment" by Mary Lee Hahn** (1st Grade, Week 6, page 74).

# WEEK 10: MORE MATTER

## ELEMENTAL
by **Jeannine Atkins**

Marguerite Perey washed test tubes that once
glowed blue. Could she discover an element?
Madame Curie, who ran the laboratory,
had found two: polonium and radium.

Most elements had already been found.
Calcium is part of bones, pearls, and chalk.
Iron is both in blood and far underground.
Oxygen and hydrogen are almost everywhere.

Would another element glow like gold,
last like lead, stink like sulfur, or glare like neon?
Every element is unique, we're told.
Arsenic poisons. Chlorine cleans.

Marguerite isolated francium, the last element
to be found in nature. Or will there be more?
Seekers and celebrants raise beakers
to all that's beyond and what came before!

**Note:** Marguerite Perey (1909–1975) worked in Paris with Marie Curie, who won two Nobel Prizes, and her daughter, Irène Joliot-Curie, who won one Nobel Prize. Both encouraged the work of talented female chemists and physicists, but it wasn't until 1962 that Marguerite Perey became the first woman to be admitted to the French Academy of Sciences.

### Take 5!

1. **Show a table of the periodic elements as a backdrop** while you read the poem aloud. One interactive link is WebElements.com.

2. Read the poem aloud again, and **challenge students to chime in on each of the element names** (*polonium, radium, calcium, iron, oxygen, hydrogen, gold, lead, sulfur, neon, arsenic, chlorine, francium*) as you read the rest of the poem aloud.

3. **Work together to learn more about Marguerite Perey and her work.** One resource is Wikipedia.org.

4. Use this poem to talk about how **matter has measurable physical properties and how those properties determine how elements are classified, changed, and used.** Explore each of the elements highlighted in the poem and research the history and applications of each.

5. Compare this poem with another about a diligent researcher by revisiting **"That Dish Thing" by Virginia Euwer Wolff** (5th Grade, Week 8, page 236). For more poems about women who have made a difference, look for *Vherses: A Celebration of Outstanding Women* by J. Patrick Lewis.

## Week 11: Force, Motion & Energy

**Fifth Grade**

### No Penguins Here
by **Michael Salinger**

It has been said
that opposites can attract
and while this may not be a fact
between your brother or sister and you
it most certainly will hold true
when magnets are the topic.

Every magnet has two poles—
one north and one south at either end.
Whether two magnets repel or attract
depends on how these poles align.
You'll get the same result every single time—
but go ahead and try it.

South pole to south pole
or north to north
 push each other away
but north and south come together—
they stick and they stay.
Because at each pole the magnetic field
is at its max
doing its job, staying on track
ensuring that—at least with magnets
opposites do attract.

### Take 5!

1. **If you have a magnet handy, now is the time to use it as your poetry prop** as you read the poem aloud.

2. Share the poem again, and **encourage students to join in on the second line (*that opposites can attract*) and last line (*opposites do attract*).**

3. **Invite students to choose one stanza to illustrate with a simple drawing or diagram.** Then arrange their illustrations next to a copy of the poem to show the variety of possible poem interpretations and visualizations.

4. Use this poem to talk about the effect of force on an object. **Use simple magnets to conduct a few quick experiments.** One helpful activity book is *Hands-On Science: Electricity and Magnets* by Sarah Angliss and Maggie Hewson.

Link this poem with **"Testing My Magnet" by Julie Larios** (1st Grade, Week 7, page 75) or **"Love Note to a Magnet" by Patricia Hubbell** (1st Grade, Week 11, page 79).

# WEEK 12: MORE FORCE, MOTION & ENERGY

**FIFTH GRADE**

## AT THE SPEED OF LIGHT
by **Shirley Smith Duke**

The sun gives off waves of energy,
hurled at the speed of light.
Particles from these waves
pass through the vacuum of space.
Waves are spaced with crests and troughs,
each kind with its own length.
Line up each one to see
waves of the electromagnetic spectrum.

Radios and phones tune in long **radio** waves.
**Micro**waves heat water in our food.
**Infrared** warms us in fires and sun's rays.
**Visible light** waves bring bright colors to our eyes.
**Ultraviolet** waves burn skin to a sore sunburn.
Cavities and broken bones show in short **X-ray** waves.
Shortest **gamma** waves arrive with lightning flashes.

Long or short, visible or invisible,
moving in different frequencies,
some reaching Earth while others are absorbed
by the protective atmosphere:
in differing wavelengths, varied in size, all
**radiant energy.**

### Take 5!

1. Before sharing this poem, **challenge students to listen for all the different kinds of energy mentioned** in this poem. Then read the poem aloud.

2. Share the poem again, and **invite students to chime in on all the words in bold** (*radio, micro, infrared, visible light, ultraviolet, X-ray, gamma, radiant energy*) while you read the rest of the poem aloud.

3. **Work together to make a simple chart** featuring each of the forms of energy cited in the poem along with examples of how each is useful.

4. **Use this poem to talk about how energy occurs in many forms** and can be observed in cycles, patterns, and systems. Explore the uses of each energy form. One useful resource is the U.S. Energy Information Administration EIA.gov/kids/energy.cfm?page=2.

5. Connect this poem with another about energy sources, **"Zapped!" by April Halprin Wayland** (3rd Grade, Week 7, page 155). For additional information, seek out the popular nonfiction book *DK Eyewitness: Energy* by Jack Challoner.

## WEEK 13: LIGHT & SOUND

**FIFTH GRADE**

### Take 5!

1. As you read this poem aloud, **tap your pen on the desk in a rhythmic way** as suggested in the poem.

2. Share the poem again, and **invite students to provide the sound effects**, first one student tapping a pen for the first four stanzas, then the whole class tapping pens throughout the rest of the poem while you read the poem aloud.

3. **Look for guidance on conducting simple sound experiments** in *Tabletop Scientist: The Science of Sound: Projects and Experiments with Music and Sound Waves* by Steve Parker.

4. **Use this poem to talk about differing forms of energy, including sound.** Identify a variety of familiar sounds at FreeSound.org and talk about how these sounds are made, recorded, and used.

5. Connect this with other poems focused on sound like **"Stopwatch" by Janet Wong** (Kindergarten, Week 8, page 36); **"Pass Me Those Ear Muffs" by Graham Denton** (2nd Grade, Week 2, page 110); or **"Sound Waves" by Michael Salinger** (3rd Grade, Week 13, page 161).

## TEACHER'S LOOK
### by Shirley Smith Duke

"Don't tap on the tables,"
Our teacher said once.
But pens click so nicely
My hand failed to note.

Taps send compressed waves
Right through the air.
The vibrations pass through,
Inside both inner ears.

She stops—Oh, the look!
Now what shall I do?
Transport to a vacuum
Where sound can't move.

She points to her notebook,
And writes down my name.
Then I vow to stop tapping
And fold my hands when . . .

The whole class starts tapping—
A symphony of sound
Fills the room with vibrations
Of more taps and noise.

The synchronized beats
Make her give up her look.
My teacher starts laughing
And closes her book.

**WEEK 14: SPACE**

**FIFTH GRADE**

### COMET HUNTER
by **Holly Thompson**

ever since he was a schoolboy
studying the Great Comet Ikeya-Seki
Yuji Hyakutake had wanted
to discover a comet
so he moved to Kagoshima
with clear mountain air
and night after night
while his wife and sons slept
he aimed his huge binoculars
at one patch of eastern sky

memorizing the stars
their exact locations
sketching positions
of constellations
hoping for a rare
blaze of ice and dust

until late one winter night
he spotted not-so-bright 1995 Y1
then the next month, near
Libra and Hydra—another

"I must be dreaming," he thought
as he sketched and photographed
measured and documented
till day dawned on the mountain

"Possible comet," he reported
then waited as top astronomers
at the National Observatory in Tokyo
checked and finally confirmed
he was the first to discover 1996 B2

soon, as the orbit neared earth,
everyone was gazing up
at that glow of ice and dust
with the long sweep of tail—
at the newly named, the truly Great
Comet Hyakutake

### Take 5!

1. **Create the backdrop for this poem by playing a portion of the video** "Comet Hyakutake: The Movie" without sound while you read the poem aloud. The video can be found at YouTube.com/watch?v=H6uCHBqfZlk (start at the 8:00 interval).

2. Share the poem again, and **invite students to chime in on the important announcement ("Possible comet")** while you read the rest of the poem aloud.

3. Work together to **research the pronunciation of unfamiliar words in the poem,** like the comet's name, "Hyakutake." One resource is PronounceHow.com.

4. **Use this poem to talk about how scientists spend years studying the sky** to discern recognizable patterns and anomalies in the natural world and among the sun, earth, and moon system. For more information and visuals, consult SolarSystem.NASA.gov.

5. Pair this poem with **"The NEO Hunters" by Juanita Havill** (4th Grade, Week 14, page 202), and share selections from *Comets, Stars, the Moon, and Mars* by Douglas Florian and from *Destination: Space* or *Comets, Meteors, and Asteroids* both by Seymour Simon.

## Week 15: Sun, Earth & Moon

**Fifth Grade**

### Take 5!

1. Use a light or window to **create a shadow to set the stage for this poem.** Then read it aloud slowly with a pause between each stanza.

2. Divide the students into four groups, **with each group joining you on reading one stanza of the poem out loud:** morning, noon, afternoon, and dusk.

3. If possible, **collect data on student shadows at three points during the day** (morning, noon, afternoon) to compare with the poem. Or try simple experiments with light and shadow based on *Tabletop Scientist: The Science of Light* by Steve Parker or HighlightsKids.com/science-experiment/trace-shadows.

4. **Use this poem to talk about the cycle of day and night** caused by the earth's rotation on its axis once approximately every 24 hours. For a video model, consult YouTube.com/watch?v=R3jr0DaV8N8.

5. Link this poem with **"Considering Copernicus" by Bobbi Katz** (3rd Grade, Week 31, page 179) and the book *Earth* by Seymour Simon.

## The Shadow Grows (and Shrinks, and Grows)
### by Laura Purdie Salas

I block sun's rays from reaching ground
as morning floods the eastern sky.
To my west—and sidewalk-bound—
I cast a soundless shadow spy.

At noon, my shadow's incomplete.
The sun above wants it to be
barely bigger than my feet:
a short and stumpy shadow me.

In afternoon, the sun goes low.
I watch my echo dance and spin.
My hazy person starts to grow—
I get to know my shadow twin.

At dusk, as far east as I see,
my legs begin to deepen, stretch.
I reach out for infinity,
my weedy, reedy shadow sketch.

## Week 16: The Water Cycle

### Ocean Engine
by **Leslie Bulion**

Sun beats,
Ocean heats,
Evaporates
where trade winds meet.
Clouds form,
Tropics storm
Heat's released
Air rises, warm.
Wind belts blow,
Currents flow,
Nutrients mix
So plankton grow,
Spinning on
This Earth we know.

But greenhouse gases slash the time it
Takes to change Earth's fragile climate.

Sun beats,
Sea *over*heats,
Polar ice
Shrinks and retreats.
Shifting rains,
Hurricanes,
Rising seas
Swamp coastal plains.
Winds squall,
Currents brawl,
Biodiversity
Numbers fall—
Which engine's running?
*It's your call!*

### Take 5!

1. If possible, **hold a globe as your poetry prop** while you read this poem aloud. Read the first long stanza in a happy voice, then use an ominous voice for the second long stanza.

2. **Invite students to read the transitional two-line couplet in the middle of the poem** while you read the rest of the poem aloud again. Pause dramatically before reading the second long stanza.

3. **Highlight some of the words in this poem that refer to Earth's various systems**: the geosphere (*plains*), biosphere (*plankton, we* [humans], *biodiversity*), hydrosphere (*ocean, sea, ice, rains, hurricanes*), and atmosphere (*winds, air, greenhouse gases*). Then talk about the poem's last line and ways that humans can keep the ocean engine running smoothly.

4. Encourage students to **discuss how the ocean supports a variety of ecosystems and organisms**, shapes landforms, and influences climate. For a useful infographic and explanation of the water cycle, visit the Education Resources of The National Oceanic and Atmospheric Administration (NOAA) at Education.NOAA.gov/ Freshwater/ Water_Cycle.html.

5. For more poems on how the sun and the ocean interact, see **"Water Round" also by Leslie Bulion** (2nd Grade, Week 16, page 124). Or look for *At the Sea Floor Café: Odd Ocean Critter Poems* also by Leslie Bulion.

## Week 17: Weather & Climate

### FIFTH GRADE

### CLIMATE VERSUS WEATHER
#### by Joan Bransfield Graham

Climate's your personality,
weather is
your mood—
a warm and sunny outlook,
with occasional
attitude.

Low pressure grumbles in with rain,
an atmospheric
pout.
High pressure smiles and saves the day,
sweeps the stormy
out.

Where are you living on this globe—
your latitudinal
position?
Location has a lot to do
with your climatic
disposition.

### Take 5!

1. Before sharing this poem, challenge students to listen for particular weather words. **Then read the poem aloud and make a list of all the words they can identify** (e.g., *warm, sunny, low pressure, rain, atmospheric, high pressure, stormy, latitudinal position, location, climatic*).

2. **Break the students into two groups**—one to join in on the low pressure lines (*Low pressure grumbles in with rain / an atmospheric / pout*) and another to read the high pressure lines (*High pressure smiles and saves the day, / sweeps the stormy / out*) while you read the rest of the poem aloud.

3. Check the weather forecast in your area on Weather.com. **It's time to discuss emergency preparedness** and review emergency procedures for school or home together.

4. Use this poem to **talk about the differences between *climate* and *weather*.** Work together to make a simple chart listing attributes and examples of each. Consult nonfiction books like *Weather* by Seymour Simon or the *DK Eyewitness Book: Weather* by Brian Cosgrove.

5. Link this poem with another about meteorology, **"Weather Map" also by Joan Bransfield Graham** (4th Grade, Week 17, page 205), and with selections from *Seed Sower, Hat Thrower: Poems about Weather* by Laura Purdie Salas and *Weather Report* by Jane Yolen.

# WEEK 18: FORCES OF NATURE

**FIFTH GRADE**

## HARBOR WAVE AT HILO: TSUNAMI SURVIVOR
### by Carole Boston Weatherford

Hawaii, April Fools' Day 1946.
Kazu and his classmates
see a strange tide suck the water
from the beach at Laupahoehoe.

Curious children scamper
to the shore. They see a rainbow—
butterfly, parrot and puffer fish, lion fish
and triggerfish—flopping on the sand.

Then, the earth rumbles, the ocean
roars and a monster wave rushes in.
The mighty wave shreds school buildings
and sweeps Kazu and his friends to sea.

The island a speck in his eye,
Kazu paddles wildly, treading water
amidst sea turtles, whitecaps, and splintery
wreckage of boats and buildings.

Tired, Kazu grasps a wooden board; makes it
his raft. Can't sleep for fear of sharks.
One long day later, a sailor saves him.
Kazu's mother calls off the funeral; fixes a feast.

**Note:** A tsunami is a series of extremely long, traveling ocean waves caused primarily by earthquakes below or near the ocean floor. Tsunami waves can move at more than 500 miles per hour and can reach heights of more than 100 feet in coastal waters. During the 1990s, more than 4,000 people were killed by ten tsunamis. Eighty percent of tsunamis occur in the Pacific Ocean; however, tsunamis also threaten coastlines off the Indian Ocean, Mediterranean Sea, Caribbean region, and even the Atlantic Ocean. This story is based on the real-life experience of Yoshikazu Murakami, a Hawaiian teenager who survived being swept to sea by the tsunami of April 1, 1946. Caused by an earthquake in the Aleutian Islands, this tsunami claimed 159 lives and destroyed boats, piers, roads, bridges, railroad tracks, and buildings.

### Take 5!

1. Before reading this poem aloud, **set the stage by showing Hawaii's location on a world map.** If possible, pinpoint the beach at Laupahoehoe in particular.

2. Share the poem again, **and invite students to join you in reading the important first and last lines out loud** (*Hawaii, April Fools' Day 1946 / Kazu's mother calls off the funeral; fixes a feast*).

3. Want to learn more about this true story? **Lead students to Hawaii's Pacific Tsunami Museum website and particularly Kazu's story at** Tsunami.org/survivors.html.

4. **Use this poem to talk about what a tsunami is like.** Challenge students to cite lines from the poem (e.g., *Then, the earth rumbles, the ocean / roars and a monster wave rushes in*). Work together to research the geological causes of these "monster" waves.

5. Connect this poem with another one about the earth's geology, **"Plates" by Ann Whitford Paul** (2nd Grade, Week 18, page 126), and look for the compelling picture book *Tsunami!* by Kimiko Kajikawa.

## Week 19: Soil & Land

### FIFTH GRADE

### SOIL INVENTORY
by **Kate Coombs**

Fifteen dissolving
veins of leaves,

fading roots,
disappearing shoots,

six wing cases, thoraxes,
legs thinner than stems,

the night earth of eleven
bugs, beasts, and birds,

a scrap of paper slowly
returning to wood,

the toe bone of a mouse,
a dried bit of worm,

tiny variegated rocks
that were boulders once,

a fragment of snail shell,
two marigold seeds,

and from a lost bracelet,
one red plastic bead.

---

### Take 5!

1. If possible, **set the stage for this poem with a container of soil that you dump out in front of you** before reading this poem aloud slowly.

2. Then read the poem aloud again, and **invite students to join in on their favorite two-line couplet describing something found in the soil** while you read the rest of the poem aloud.

3. **Talk about the role of composting in creating fertilizer for soil.** Look for educator resources at the CompostingCouncil.org.

4. Scientists study the properties of soils including color and texture, capacity to retain water, and ability to support the growth of plants. **Discuss the differences between *soil* and *dirt*** using external resources such as Soils4kids.org.

5. Compare this poem with **"Magic Show" by Juanita Havill** (1st Grade, Week 19, page 87) and follow up with selections from *Footprints on the Roof: Poems about the Earth* by Marilyn Singer.

## WEEK 20: NATURAL RESOURCES

### RESOURCES RULE!
by **Susan Blackaby**

As stewards of the biosphere,
it is our job while we are here
to nurture and rejuvenate
the resources it takes to make
the things we use to meet our needs,
our comforts and necessities.

Ore, minerals, and oil that hide
in veins and caverns deep inside
Earth's firm and fragile outer crust
are quarried, mined, and drilled by us.
These raw reserves may disappear
without responsibility and care.

Preserve the precious watershed—
the run-off and the river bed.
Wise use in cities, farms, and towns
keeps water pure upstream and down,
so waterways run clear and clean
along their course from spring to sea.

Conserve the forests, lush and green,
from humus up to canopy.
Protect the woodland denizens
that need the trees for nests and dens.
Every time you hike around it,
leave Earth better than you found it.

The rich resources on our planet
can be things we take for granted,
but they'll be in short supply
if globally we don't comply
with simple rules we all respect:
reserve, preserve, conserve, protect.

### Take 5!

1. **Project beautiful images of our biosphere as a backdrop** for reading this poem aloud. One fun source is Calm.com.

2. Share the poem again and **encourage students to join you in shouting out the final line** *reserve, preserve, conserve, protect* together.

3. **Talk with students about organizations that focus on conservation,** like the World Wildlife Federation (Panda.org).

4. Use this poem to talk about how **scientists study and identify alternative energy resources** such as wind, solar, hydroelectric, geothermal, and biofuels. Challenge students to cite examples of various energy sources using examples from the poem and beyond. For more on renewable and nonrenewable energy sources go to EIA.gov/kids/energy.cfm?page=2.

5. Follow up with the poem **"Fossil Fuels" by Janet Wong** (2nd Grade, Week 20, page 128) and selections from *Earthways, Earthwise: Poems on Conservation* edited by Judith Nicholls.

## Week 21: Ecosystems

**Fifth Grade**

### Take 5!

1. To set the stage for this poem, **place a green plant in front of you** before reading the poem aloud.

2. Read the poem aloud again, and **invite students to join in on the important last line of the first stanza** (*plants are number one*) while you read the rest of the poem aloud.

3. **Work together to sketch a diagram of the cycle of photosynthesis** (clean air, water, green plant, light). Need help? Go to PhotosynthesisForKids.com.

4. Use this poem to talk about how the flow of energy is derived from the sun, is used by producers to create their own food, and is then transferred through a food chain and food web to consumers and decomposers. **Discuss the significance of the carbon dioxide-oxygen cycle** to the survival of plants and animals. One helpful model is at YouTube.com/watch?v=j4Ah1jercqQ.

5. Share a related poem, **"We Need Green Seaweed!" by Margarita Engle** (3rd Grade, Week 16, page 164), and look for more poetry in Juanita Havill's *I Heard It from Alice Zucchini: Poems About the Garden*.

## Cool Food for Thought
### by Sara Holbrook

Plants! The original solar panels,
whether swaying or standing still
transfer
blue and red wavelengths of sun
into 30 shades of green
known as chlorophyll.
Whether you pluck your food from a tree
or eat it on a bun,
of all the lion-human-chicken
links in the food chain,
plants are number one.

But plants not only feed our stomachs,
they also scrub the air,
converting carbon emissions
into the oxygen we share.
Sustaining plants are an army of organisms,
7 billion in every teaspoon of healthy soil,
together they feed us and cool the atmosphere
so we won't starve
or start to boil.

## Week 22: Adaptations & Traits

**Fifth Grade**

### What Is a Foot?
by **Jane Yolen**

You will find a foot at the end of your limb,
Where you might wear a fin when you go for a swim.
It's got segments galore, it's got bones by the dozens,
And the bones have more bones, who are all sort of cousins.

As for animal feet, there's a soft foot, or paw,
That ends in strong nails, and is often called claw.
But others have hard feet, a hoof as we say.
And that is a feat of foot facts for today.

---

### Take 5!

1. **As you read this poem aloud slowly, extend, bend, rotate, and tap your foot,** as a poem "demonstration."

2. Share the poem again, and **invite students to join in on the important last line** (*And that is a feat of foot facts for today*) while you read the rest of the poem aloud.

3. **Make a list of "foot words" used in the poem** and identify animals with examples of each (*foot, limb, fin, segments, bones, paw, nails, claw, hoof*).

4. Use this poem to talk about all different kinds of feet. **Work together to research images of each of the feet featured in the poem.** Compare the structures and functions of different species that help them live, adapt, and survive, such as hooves on prairie animals or webbed feet on aquatic animals.

5. For more poems on body parts, look for **"Froggy" by Charles Waters** (4th Grade, Week 6, page 194) and *The Blood-Hungry Spleen and Other Poems about Our Parts* by Allan Wolf.

## WEEK 23: CYCLES

**FIFTH GRADE**

### Take 5!

1. **Play cicada sounds in the background as you read this poem aloud**. One source is FreeSound.org.

2. Read this poem again, and **encourage students to chime in on the line *What do you want from me?*** while you read the rest of the poem aloud.

3. Since this is a bilingual poem, it's the perfect opportunity to **invite anyone who speaks Spanish (in your class, school, or community) to join you by reading the Spanish poem** before or after you read the poem in English. If necessary, you could use VoiceThread to make a long-distance recording of the Spanish reading with a Spanish speaker in advance.

4. Use this poem to **talk about the life cycle of the cicada and its place in the environment.** One useful resource is CicadaMania.com.

5. Connect this poem with **"Becoming Butterflies" by Jeannine Atkins** (2nd Grade, Week 23, page 131) and selections from *Insectlopedia* by Douglas Florian and *Cicadas!: Strange and Wonderful* by Laurence Pringle.

## CICADA
### by **Guadalupe Garcia McCall**

I dug my toes into the dirt
and felt a tickle at my heel.
A pretty nymph had crept up there
and made me jump and squeal.
"What do you want from me?"
I asked the young cicada.
"I need to climb up on that tree,
to feel the sun, to have some fun,
to shed my skin, and rest and sing.
I need some time to think,
to shake the daze from all that sleep,
and dust off my brand new wings."
So I let her climb upon my finger
and placed her gently on the bark
that she may crawl up to the top
and someday soon enjoy the park.

## CHICHARRA
### por **Guadalupe Garcia McCall**

Metí los dedos del pie en la tierra
y sentí cosquillas en el talón.
Una ninfa bella se había trepado allí
y me hizo brincar y gritar.
—¿Qué quieres tú de mí?—
le pregunte a la joven chicharra.
—Necesito subirme a ese árbol,
sentir el sol, disfrutar,
mudar de piel, descansar y cantar.
Necesito un poco tiempo para pensar,
sacudir el aturdimiento de tanto sueño
y desempolvar mis alas nuevecitas.—
Por eso la dejé treparse a mi dedo
y la puse sobre la corteza cuidadosamente
para que gateara hasta la copa del árbol
y algún día en breve gozara del parque.

## WEEK 24: PATTERNS

**FIFTH GRADE**

### PATTERNS IN NATURE : NATURE IN PATTERNS
by **Shirley Smith Duke**

| | |
|---|---|
| patterns in nature | nature in patterns |
| bilateral symmetry | symmetry bilateral |
| middle a line | line a middle |
| halves in two | two in halves |
| matching exactly | exactly matching |
| one makes another | another makes one |
| identical images | images identical |
| —exactly alike— | —alike exactly— |
| mirror opposites | opposites mirror |

### Take 5!

1. **Create a backdrop for this poem by showing images of bilateral symmetry**, such as butterflies or beetles. Then read the poem aloud pausing after each phrase.

2. As you read the poem aloud again, **challenge students to read the right-hand column of lines** while you read the left-hand column.

3. Using the visuals at Photography.NationalGeographic.com, **explore the variety of patterns in nature**—in reflections, landscapes, flowers, rocks, animals, and so on.

4. **Lead students in discussing the principle of symmetry** in the natural world. Research organisms that demonstrate bilateral or radial symmetry and identify those lines of symmetry.

5. Connect this with the poem **"Rings Not Letters" by Juanita Havill** (2nd Grade, Week 24, page 132) and with selections from *Forest Has a Song* by Amy Ludwig VanDerwater and *Swirl by Swirl: Spirals in Nature* by Joyce Sidman.

## Week 25: Human Body

### FIFTH GRADE

### Take 5!

1. As you read this poem aloud, pause between each stanza and **move your arm and elbow as described in the poem.**

2. Read the poem again, and **invite students to join you in the movement suggested by the poem** (bending elbows, touching toes, brushing hair, blowing nose).

3. **Work together to draw a simple diagram of the elbow** complete with hand, arm, and shoulder. Label according to the poem (A, B, C, vertex, endpoints).

4. **Lead students in a discussion of bones, muscles, and joints** and how they function in human anatomy. Use the teacher's guide found at KidsHealth.org at Classroom.KidsHealth.org/3to5/body/parts/bones.pdf.

5. Compare human elbows with the joints of other animals and revisit **"What Is a Foot?" by Jane Yolen** (5th Grade, Week 22, page 250). Also seek out *The Blood-Hungry Spleen and Other Poems about Our Parts* by Allan Wolf.

### LET ME JOIN YOU
by **Heidi Bee Roemer**

Your
elbow
isn't cute.
It is wrinkly,
red, and rough. But just
try to live without it!
Simple tasks would sure be tough.

Your
elbow
**joint** is a
**vertex.** Mark it
with a B. Then draw
an A upon your hand.
Now let's tag your shoulder C.

With-
out your
elbow joint,
I wonder, how
would you touch your toes?
How would you brush your hair?
And how would you blow your nose?

Your
elbow
isn't cute,
but that **vertex**
labeled B, keeps your
useful parts connected,
like your **endpoints** A and C.

*The Poetry Friday Anthology for Science*

## WEEK 26: KITCHEN SCIENCE

**FIFTH GRADE**

### THIRSTY MEASURES
by **Heidi Bee Roemer**

I pour some juice into my **cup**.
8 fluid ounces I drink up.
That's 16 tablespoons. Good stuff!
But the juice is not enough.

I mix a **pint** of lemonade,
the pink kind that my grandma made.
That's 16 ounces, 2 cold cups—
The lemonade is not enough.

I chug a **quart** of chocolate milk
and not one drop of milk is spilt.
That's 32 ounces, 4 cold cups,
2 pints of milk. Still . . . not enough.

I swig a **gallon** of iced tea.
That's 16 cups of tea for me.
I reach my peak capacity . . .
8 pints, 4 quarts, all gone. Oh no—
Excuse me, please. I gotta go!

### Take 5!

1. If possible, add some fun to sharing this poem with a poetry prop—**show a measuring cup** before reading the poem aloud.

2. Share the poem again, and **invite students to chime in on all the numbers in the poem (*8, 16, 2, 32, 4*)** while you read the rest of the poem aloud.

3. **Survey students on their favorite beverages** (juice, lemonade, chocolate milk, iced tea, and so on) and make a simple chart documenting the results. Research recommendations on healthy eating at Fitness.gov/eat-healthy/how-to-eat-healthy/.

4. Scientists frequently use measurement as part of their investigations and experiments. **Lead students in discussing the measurement equivalencies presented in this poem.** Collaborate to create a simple poster or chart to review those measurements. One resource is Star.Spsk12.net/math/3/Gallonman.ppt.

5. Connect this poem with **"Liquids Can't Contain Themselves" also by Heidi Bee Roemer** (2nd Grade, Week 10, page 118) and selections from *Marvelous Math: A Book of Poems* edited by Lee Bennett Hopkins or *Edgar Allan Poe's Pie: Math Puzzlers in Classic Poems* by J. Patrick Lewis.

## Week 27: Video Technology

### Frames Per Second (fps)
by **Janet Wong**

I'm trying to do animation
for my own video game.
I need to divide each action
into pieces, frame by frame.
I wonder: how many frames?
I don't want it to look rough.
I do a search on the Web
and read some shocking stuff:
most video games use
thirty frames per second (fps)—
some use much, much more,
some a little less.
So now I have an idea,
a smarter, better one:
I'll outsource all the work—
and I'll just play for fun!

### Take 5!

1. After reading this poem aloud, **use Voki.com to create a simple avatar for yourself** and encourage the students to create their own avatars, too.

2. Read the poem aloud again, and **invite students to chime in when the title of the poem appears within the poem** (*Frames Per Second (fps)*).

3. Collaborate with students to **create a quick glog, a digital interactive poster (using Glogster.com), pulling together images and key words from the poem** in a new, visual representation of the poem's theme. Show the students the choices of text, fonts, color, graphics, and even animation, if possible, while you input those items and create the finished product.

4. **Talk with students about what a game designer does**—working with computers, programs, and animation to create new games. Explain terms like *frame* and talk about how technology tools are used to solve problems and create innovations.

5. Match this poem with **"Virtual Adventure" by Renée M. LaTulippe** (4th Grade, Week 27, page 215), and check out *The Crazy Careers of Video Game Designers* by Arie Kaplan.

# WEEK 28: MACHINES

## SODA MACHINE BITE
by **Jacqueline Jules**

Five rows of plastic bottles,
lined up in bright colors
like alphabet blocks.
All I have to do is
feed two bills into
a metal mouth with rollers
and press button
A5 for lemon-lime.
Wait!
It's B5 for lemon-lime,
and A5 for cola.
Too late!
The dispensing coil
punches the wrong bottle
like a bulldozer
down to the drawer by my knees.
That's not the drink I wanted,
but my money has been scanned
by optical sensors and swallowed.
This is the drink I get.
Might as well unscrew the top
and enjoy!

### Take 5!

1. Start this poem with a **poetry prop—a soft drink or water bottle—** placed in front of you. Then read the poem aloud, stopping briefly wherever punctuation suggests a pause.

2. Share the poem again, and **invite students to chime in on all the numbers and abbreviations** in the poem (*five, two, A5, B5*) while you read the rest of the poem aloud.

3. For discussion: **Should vending machines provide only healthy drink choices?**

4. Use this poem to talk about the various components of a soda machine (as described here). **Challenge students to draw and label a model** that shows how a soda dispensing machine works. Consult David Macaulay's *The New Way Things Work* and the homemade video at YouTube.com/watch?v=Smbyr9eZcd0.

5. Link this poem with another about everyday machines, **"Metal Monster" by X. J. Kennedy** (Kindergarten, Week 28, page 56), and look for easy experiments to do in *Simple Machines* by Deborah Hodge.

## Week 29: Building Things

### FIFTH GRADE

### Take 5!

1. **Set the stage for this poem by projecting a photograph of the pyramids** at Giza. One source is NationalGeographic.com/pyramids/khufu.html. Then read the poem aloud, pausing briefly after each stanza.

2. Read the poem again, and **encourage students to join in on the lines beginning with *No*** (*no steel, / no great machines, / no software help; No pulleys pulled, no big cranes raised*) while you read the rest aloud.

3. Do a bit of quick **collaborative research on the pyramids at Giza.** Use the link DiscoveringEgypt.com/pyramid3.htm.

4. **Work together to make a checklist of all the tools for building** that are itemized in this poem (e.g., steel, machines, computers, *wooden "tool of knowing," palm leaf spine,* pulleys, cranes, ropes, levers, iron beam). Then talk about which were used in building the pyramids and which were not. For more information, look for *Pyramid* by David Macaulay or *Eyewitness: Pyramid* by James Putnam.

5. For a modern look at the tools we use for building, seek out **"Computer Models" by Janet Wong** (4th Grade, Week 8, page 196) and follow up with *Monumental Verses* by J. Patrick Lewis.

## The Great Pyramid of Giza
### (completed around 2570 BC)
by **Laura Purdie Salas**

Four thousand sandy years ago,
Egyptians built a sacred site.

They had
no steel,
no great machines,
no software help
to get it right.

They used a wooden "tool of knowing"
to read the sky, keep stars aligned.
They aimed a fragile palm leaf's spine
to keep all angles well-defined.

For 20 years, the workers toiled,
hour by hour and block by block.
They quarried limestone in the sun
and heaved out mighty granite rock.

No pulleys pulled, no big cranes raised.
With papyrus ropes, they dragged each stone—
two tons—then levered into place…
We think—so much is still unknown!

Made of rock, with wood and sweat,
built without one iron beam.
Giza held the record height
for centuries.

Now, that's extreme!

*The Poetry Friday Anthology for Science*

## Week 30: Science Fair

### Tell It to the Court
by **Janet Wong**

I didn't mean to copy.
I would never plagiarize.
The words
somehow just . . .
went
straight from the screen
into my eyes
and down
into my fingers
and into my report.

Please-oh-please
don't tell me
to "tell it to the court"!

### Take 5!

1. After reading this poem aloud, **discuss the meaning of the word *plagiarize*** (to take the work or ideas of someone else and claim them as one's own).

2. Share the poem again, and **invite students to chime in when the title of the poem appears within the poem** (*Tell It to the Court*) while you read the rest of the poem aloud.

3. **Model how to use external sources of information**, take notes, cite sources, and frame ideas in new, individual ways. One excellent resource is KidsHealth.org/kid/feeling/school/plagiarism.html.

4. Use this poem to **talk about how important ethical behavior is to scientists.** We communicate findings and conclusions in both written and verbal forms and constantly evaluate the accuracy of information. What if we relied on information that was wrong or claimed ideas that were not our own?

5. Connect this poem with another that explores the concept of truth, **"Wiki Alert" by Debbie Levy** (3rd Grade, Week 33, page 181).

## Week 31: Famous Scientists

### Take 5!

1. Set the stage for this poem by showing a map locating the Zhejiang Province of China and talking about the timeframe for this subject (1031-1095). **Encourage students to visualize that time and place** as you read the poem aloud.

2. Share this poem again, and **invite students to join in on the pivotal line** (*But people did not want to hear these things*) while you read the rest of the poem aloud.

3. For discussion: *Why do people sometimes ignore the claims of science?*

4. **Work together to research Shen Kuo's many contributions to science,** starting with those listed in the poem. One helpful resource is ChinaCulture.org/gb/en_madeinchina/2003-09/24/content_72415.htm.

5. For another poem about climate change, revisit **"Ocean Engine" by Leslie Bulion** (5th Grade, Week 16, page 244), or for a poem about magnetic poles, seek out **"No Penguins Here" by Michael Salinger** (5th Grade, Week 11, page 239).

## Shen Kuo
### (1031–1095)
by **Janet Wong**

Almost a thousand years ago
a Chinese scientist named Shen Kuo,
geologist-cartographer-astronomer-engineer,
discovered fossilized shells
hundreds of miles inland
that made it clear the shoreline had moved.
Petrified bamboo convinced him
that climate change was happening.
But people did not want to hear these things.
Instead he became known for the idea
that true north is not magnetic north.
His magnetic needle compass was worth
spices, gold, jewels—even a giraffe—
as explorers later sailed to Africa and back.

Climate change and the shifting sea:
who would choose such mundane news
over promises of spices, gold, and jewels?

# WEEK 32: MORE FAMOUS SCIENTISTS

## I WILL BE A CHEMIST: MARIO JOSÉ MOLINA
### by Alma Flor Ada

Only a drop of water
but looking under the microscope
I see things that move inside that very drop.
My aunt Esther has given me a chemistry set.
She says that everything—water, air, earth,
the trunks of trees and our own skin—
is made of small particles we cannot see.
She explains that even these molecules
are made of chemical elements;
just around a hundred elements
combine to make all that exists.
I have started today
in my simple lab in the old unused bathroom
to study these elements.
I will know the secrets of the universe.

**Note:** Mario José Molina, born in Mexico in 1943, and a resident of Mexico and California, won the Nobel Prize in 1995 for his contribution to understanding the ozone layer. He continues his work to solve pollution problems and to study atmospheric particles and their effect on clouds and climate.

## VOY A SER QUÍMICO: MARIO JOSÉ MOLINA
### por Alma Flor Ada

Sólo una gota de agua
pero mirándola con el microscopio
veo cosas moviéndose en ella.
Tía Esther me ha regalado un juego de química.
Dice que todo—el agua, el aire, la tierra,
los troncos de los árboles y hasta nuestra piel—
está hecho de pequeñas partículas que no podemos ver.
Explica que las moléculas
están formadas por elementos químicos.
Apenas unos cien elementos
se combinan para crear todo lo que existe.
He empezado hoy
en mi simple laboratorio en el viejo cuarto de baño
a estudiar estos elementos.
Voy a saber los secretos del universo.

**Nota:** Mario José Molina, nació en México en 1943 y ha vivido en México y en California. En 1955 recibió el Premio Nobel por ayudar a comprender la capa de ozono. Continúa estudiando cómo resolver problemas de polución y el efecto de las partículas atmosférica en las nubes y el clima.

### Take 5!

1. Once again, **display a table of the periodic elements as a backdrop** while you read the poem aloud. Use the link WebElements.com.

2. Since this is also a bilingual poem, it's the perfect opportunity to **invite anyone who speaks Spanish (in your class, school, or community) to join you by reading the Spanish poem** before or after you read the poem in English.

3. **Work together to research the life and work of Mario José Molina.** Consider his place in the history of science and as a Nobel Prize winner for his contributions to the field of chemistry. Consult NobelPrize.org/nobel_prizes/chemistry/laureates/1995/molina-bio.html.

4. **Use this poem to talk about water,** its formula ($H_2O$), and its unique role in the life cycle of plant and animal life. Explore the resources of Chem4kids.com.

5. Link this poem with another about studying chemical elements by revisiting **"Elemental" by Jeannine Atkins** (5th Grade, Week 10, page 238) and seek out the nonfiction book *The Elements: A Visual Exploration of Every Known Atom in the Universe* by Theodore Gray.

## Week 33: Computers

### Printing, Pressed Beyond Words . . .
by **Robyn Hood Black**

Our printers today are still evolving.
So many projects—and problems they're solving!

In layers of plastic, a virtual mold:
printers are spitting out things you can hold.

These 3-D devices can also print gels,
stacking amazing assortments of cells.

Need a blood vessel? An organ, an ear?
Bioprinting is real—bioprinting is here!

### Take 5!

1. Read this poem aloud, slowly and clearly. **Follow up with a short video demonstrating the possibilities of bioprinting at** YouTube.com/watch?v=9D749wZSlb0.

2. Then share this poem again, **inviting students to say the powerful last line together** while you read the rest of the poem aloud.

3. If a bioprinter can create 3-D human tissue, **challenge students to consider what other kinds of machines they can imagine** might exist in the future.

4. Use this poem to talk about how **scientists use technology tools for solving problems.** Explore making 3-D papercrafts with plain printers; this includes paper models, puzzles, etc. Search Papertoys.com for downloadable templates.

5. For more poems about imaginative inventions, look for **"The Engineer" by Stephanie Calmenson** (1st Grade, Week 33, page 101) **or "Da Vinci Did It!" by Renée M. LaTulippe** (1st Grade, Week 31, page 99).

## WEEK 34: SCIENCE CAREERS

**FIFTH GRADE**

### A NEW DINOSAUR
by **Marilyn Nelson**

A newly discovered German shepherd-sized dinosaur
provides a wealth of new information on the evolution of
bone-headedness.
The dinosaur, identified from fossilized bone fragments
unearthed in Alberta, Canada,
is one of the earliest pachycephalosaur specimens
ever unearthed. Pachycephalosaurus
(from Greek *pachys-/παχυς-* "thick,"
*kephale/κεφαλη* "head" and *sauros/σαυρος* "lizard"),
lived during the Late Cretaceous Period
in what is now North America.
Paleontologists have named the new creature
*Acrotholus audeti*: its genus *Acrotholus*
("thick dome"), its unique species called *Audeti*
because it was found on a ranch owned by Roy Audet.
*Acrotholus audeti*—"Audi" for short—
roamed about 85 million years ago.
She walked or ran on long hind legs,
lived in herds, defended herself with head-butting.

Or so says Science, continuing Adam's task
of naming the animals. Naming, understanding,
paying due awe to each trace of the evidence
of Papa Creation's unfolding, evolving design,
which is denied by the pachycephalic now among us.

### Take 5!

1. **Display an image of a pachycephalosaurus in the background** as you read this poem aloud slowly. One source is Animals.NationalGeographic.com.

2. Share the poem again, **and invite students to chime in on the dinosaur name,** *pachycephalosaur* **and** *Audi for short* while you read the rest of the poem aloud.

3. Explore the story behind the story of this poem with the video at YouTube.com/watch?v=wxjnQEhpJvU. **Talk about how there are still discoveries to be made about dinosaurs.**

4. **Use this poem to talk about how scientists are constantly examining all sides of scientific evidence,** using empirical data, logical reasoning, and experimental and observational testing. To review geological eras, check out the link UCMP.Berkeley.edu/help/timeform.php.

5. Contrast this poem with another poem about prehistoric creatures, **"Trilobite" by Mary Ann Hoberman** (3rd Grade, Week 19, page 167). Also look for selections from *Dinothesaurus* by Douglas Florian and *Tyrannosaurus Was a Beast: Dinosaur Poems* by Jack Prelutsky.

## Week 35: Future Challenges

**Fifth Grade**

### Titan in Man's Seaweed
**(West Indian Manatees)**
by **Michael J. Rosen**

Gargantuan sirenian,
you lone marine mammalian
who's totally vegetarian,
you sea cow some have called a mermaid,
your trusting nature has been betrayed
because you never were afraid
of motorboats or nets or oceans
fouled by waste that evolution
never meant for your slow motion.
Almost extinct—how can that be?
Is it our need to eat or vanity
or inhumanity, oh manatee?

**Note:** The title is an anagram of the animal's name.

### Take 5!

1. As a backdrop for this poem, **display an image or video of a manatee**. One source is Animals.NationalGeographic.com. Then read this poem aloud slowly.

2. Share the poem again, and **invite students to join in on the important last line** (*or inhumanity, oh manatee*) while you read the rest of the poem aloud.

3. Conduct a bit of quick collaborative research to **learn more about West Indian manatees**. One resource is SaveTheManatee.org.

4. This poem can **launch a discussion about endangered animals** and how changes in ecosystems affect the life cycle of all inter-dependent organisms.

5. Compare this poem with **"The Lament of Lonesome George" by Jane Yolen** (2nd Grade, Week 35, page 143), and seek out nonfiction books for background information, like *The Manatee Scientists: Saving Vulnerable Species* by Peter Lourie.

## WEEK 36: FUTURE DREAMS

### FIFTH GRADE

### MY WRISTROBOT PACK
by **Carmen Tafolla**

I put my WristRobot on my wrist every day
As soon as it wakes me with my breakfast tray.
It lays out my clothes, all clean and pressed,
Checks out the forecast, and helps me get dressed.
Then it reminds me that my hair's sticking up,
Makes my bed, packs my lunch, and refills my cup.

It sets all the vectors to beam me to school
And makes sure I travel through Rome, Istanbul,
Paris, and Bogotá on my way,
To guarantee I have an interesting day.
With Laser-Lev games and Photoelectronic Tag,
I can play with my friends wherever they're at.

I'd NEVER be without one. I'll always have it near.
—Wait a minute! Where am I? What's happening here?
Whose messy bed is this? Why no Laser-Lev games?
Mom, is this MY home? Was this all a dream?
NO WristRobot Pack?? They don't even EXIST?!!
...
Well then, I'll just make one—it's first on my list!

Sign me up for Physics and Electro-robotics.
I need Laser Science and Transmitter-crionics.
I want to study Electromagnetic Levitation,
Bilocation Engineering, Locomotion Actuation.
I'm really missing my old WristRobot Pack—
I'll invent it and THEN, get my DREAM life back!

### Take 5!

1. If time allows, **print and cut out a picture of a small robot and make your own prototype of a "WristRobot Pack."** Incorporate it into your reading of the poem, pointing, gesturing, and pantomiming each action in the poem as if you have a WristRobot Pack.

2. Before reading the poem aloud again, **invite students to choose their favorite stanza:** Stanza 1 that helps you get ready, Stanza 2 that imagines travel and games, Stanza 3 with the surprise twist, or Stanza 4 with the exciting conclusion. Then read the poem aloud together with students joining in on their favorite stanza only.

3. For discussion: *What would you like an (imaginary) WristRobot Pack to do for you?*

4. This poem encourages students to think creatively and imagine developing new digital projects. **Talk with students about what the poet imagines here that a *WristRobot Pack* can do. Invite them to forecast what other inventions they might imagine** based on the trends and possibilities they see around them.

5. Compare this imagined robot with the one described in **"My Robot" by David L. Harrison** (2nd Grade, Week 29, page 137).

THE POETRY FRIDAY ANTHOLOGY FOR SCIENCE

## SCIENCE
### by James Carter

The birth of a star
the beat of a heart

The arc of an hour
the bee and the flower

The texture of sound
the earth spinning 'round

The river in flood
the nature of blood

The future in space
for our human race

Now that's what I call science

# THREE FUN AND EASY WAYS TO CELEBRATE SCIENCE POETRY

"Poetry Celebrations" give your students something to look forward to and can provide opportunities for child participation individually and as a community. For hundreds of additional suggestions and ideas, consult *The Poetry Teacher's Book of Lists* by Sylvia Vardell.

1. If **audio announcements** are made on a regular basis, encourage children to volunteer to read aloud a poem of their choice (and to try sound effects or choral reading).

2. **Read poems for birthdays** (e.g., favorite poems of the birthday child). Invite families to donate poetry books in honor of the birthday child. Tape the poem reading so students and families can have a record of their young voices as a keepsake.

3. **Share a poem to commemorate landmark events** like Earth Day or for special occasions like Open House or the Science Fair. Choose poems from an e-book (such as the e-book version of THIS book) and project the poem for all to see during the event.

# More Poetry Resources

"Student: Dr. Einstein, aren't these the same questions as last year's [physics] exam?

Dr. Einstein: Yes, but **this year the answers are different.**"

            ❧ Albert Einstein ☙

# Building Your Own Poetry Library

How do we identify which poetry books are the best for children or most useful in the K-5 classroom? One of the best places to begin is by looking at poetry award winners.

The **Children's Poet Laureate** (CPL) was established by the Poetry Foundation in 2006 to raise awareness of the fact that children have a natural receptivity to poetry and are its most appreciative audience, especially when poems are written specifically for them. The Children's Poet Laureate serves as a consultant to the Foundation and gives public readings. The first CPL was Jack Prelutsky, followed by Mary Ann Hoberman, J. Patrick Lewis, and Kenn Nesbitt.

Another major award for poetry for children is **the National Council of Teachers of English (NCTE) Award for Excellence in Poetry for Children**, given to a poet for her or his entire body of work in writing or anthologizing poetry for children. Several of the winners are included in this anthology: Arnold Adoff, X.J. Kennedy, Nikki Grimes, J. Patrick Lewis, and Joyce Sidman. Any book of poetry by one of these award winners will be worthwhile.

> **Your Poetry Checklist**
>
> ☑ Highlight poetry books on the chalk rail, a red wagon, or a table
>
> ☑ Seek out poetry books from diverse perspectives
>
> ☑ Link poems with picture books, novels, and nonfiction
>
> ☑ Connect children's poetry with science and mathematics
>
> ☑ Use technology to share poems, respond to poems, and expand upon poems
>
> ☑ **Tell your colleagues about Poetry Friday!**

Other prominent awards include the **Lee Bennett Hopkins Award** for Children's Poetry, which is presented annually to an American poet or anthologist for the most outstanding new book of children's poetry published in the previous year; the **Claudia Lewis Award,** given by Bank Street College for the best poetry book of the year; and **The Lion and the Unicorn Award** for Excellence in North American Poetry for the best poetry book published in either the U.S. or Canada. A detailed listing of major poetry awards, past winners, and useful award-related website links can be found in *The Poetry Teacher's Book of Lists* by Sylvia Vardell.

*The Poetry Teacher's Book of Lists* also offers input on selecting poetry for young people ages 0-18. It contains 155 different lists and cites nearly 1500 poetry books in a variety of categories, including:

- Poetry Awards and "Best" Lists
- Seasonal and Holiday Poetry Booklists (Valentine's, Earth Day, the seasons, spring, etc.)
- Multicultural and International Poetry Booklists (such as African American or bilingual poetry books)

- Poetry Booklists Across the Curriculum (animals, dinosaurs, science, space, weather, time, math, etc.)
- Strategies for Creating a Poetry-Friendly Environment (poetry displays and quotes, lesson plan tips, a poetry scavenger hunt, poet birthdays)
- Strategies for Sharing and Responding to Poetry Out Loud (poetry performance tips, assessment rubrics, discussion prompts)
- Strategies for Teaching Poetry Writing (lists of poetry written by children, lists of poem forms, writers' checklists)
- General Poetry Teaching Resources (poetry websites and blogs, poetry text sets, reference tools)

If you are looking for poetry books for Earth Day or poems for a unit on insects and bugs, for example, you'll find lists for each of those and more.

## Poetry Books for Science

A brief look at the poetry shelves will lead one to discover many poems that connect with science, such as collections devoted to **animals, weather, seasons, space, dinosaurs,** and **time,** to name a few subjects featured in lists in *The Poetry Teacher's Book of Lists* by Sylvia Vardell. Here are some science poetry books by contributors to *The Poetry Friday Anthology for Science*. Check their personal websites for recently released titles.

Blackaby, Susan. *Nest, Nook & Cranny*
Bruchac, Joseph. *The Earth under Sky Bear's Feet: Native American Poems of the Land*
Bruchac, Joseph. *Thirteen Moons on Turtle's Back: A Native American Year of Moons*
Bulion, Leslie. *At the Sea Floor Café: Odd Ocean Critter Poems*
Bulion, Leslie. *Hey There, Stink Bug!*
Coombs, Kate. *Water Sings Blue: Ocean Poems*
Dotlich, Rebecca. *What Is Science?*
Florian, Douglas. *Comets, Stars, the Moon, and Mars*
Florian, Douglas. *Dinothesaurus*
Florian, Douglas. *UnBEElievables: Honeybee Poems and Paintings*
Franco, Betsy. *Bees, Snails & Peacock Tails: Patterns and Shapes . . . Naturally*
Gerber, Carole. *Seeds, Bees, Butterflies, and More! Poems for Two Voices*
Graham, Joan Bransfield. *Flicker Flash*
Graham, Joan Bransfield. *Splish Splash*
Harley, Avis. *Sea Stars: Saltwater Poems*
Harley, Avis. *The Monarch's Progress: Poems with Wings*
Harrison, David. L. *Bugs: Poems about Creeping Things*
Havill, Juanita. *I Heard It from Alice Zucchini: Poems About the Garden*
Hubbell, Patricia. *Earthmates: Poems*
Katz, Bobbi. *Trailblazers: Poems of Exploration*
Larios, Julie. *Yellow Elephant: A Bright Bestiary*

Lewis, J. Patrick, Ed. *The National Geographic Book of Animal Poetry*
Lewis, J. Patrick. *Galileo's Universe*
Lewis, J. Patrick. *Scien-trickery: Riddles in Science*
Mordhorst, Heidi. *Pumpkin Butterfly: Poems from the Other Side of Nature*
Rosen, Michael J. *The Cuckoo's Haiku and Other Birding Poems*
Ruddell, Deborah. *A Whiff of Pine, A Hint of Skunk*
Ruddell, Deborah. *Today at the Bluebird Café*
Salas, Laura Purdie. *And Then There Were Eight: Poems about Space*
Salas, Laura Purdie. *Chatter, Sing, Roar, Buzz: Poems about the Rain Forest*
Salas, Laura Purdie. *Seed Sower, Hat Thrower: Poems about Weather*
Sidman, Joyce. *Dark Emperor and Other Poems of the Night*
Sidman, Joyce. *Eureka! Poems about Inventors*
Sidman, Joyce. *Ubiquitous: Celebrating Nature's Survivors*
Singer, Marilyn. *A Strange Place to Call Home*
Singer, Marilyn. *Footprints on the Roof: Poems about the Earth*
Singer, Marilyn. *How to Cross a Pond: Poems about Water*
Spinelli, Eileen. *Polar Bear, Arctic Hare: Poems of the Frozen North*
VanDerwater, Amy Ludwig. *Forest Has a Song*
Weatherford, Carole Boston. *I, Matthew Henson: Polar Explorer*
Wolf, Allan. *The Blood-Hungry Spleen and Other Poems about Our Parts*
Wong, Janet. *Once Upon a Tiger: New Beginnings for Endangered Animals*
Yolen, Jane. *Birds of a Feather*
Yolen, Jane. *Bug Off! Creepy Crawly Poems*
Yolen, Jane. *Least Things: Poems about Small Natures*

## Outstanding Science Trade Books for Students K-12

And don't forget to check out the annual list of "Outstanding Science Trade Books for Students K-12" selected by the National Science Teachers Association in collaboration with the Children's Book Council. This list typically includes one or two new books of science-themed poetry (as well as science-rich literature in other genres) every year. For more information and previous book lists, go to NSTA.org/ostb.

## Children's Poetry Websites and Blogs

As we look for new places for poetry to pop up, you can be sure that this includes the Internet. There are several hundred websites and blogs that make poems available; these often include audio and video recordings of poets reading their poems and/or biographical information about poets, too. A comprehensive list of poetry websites and blogs can be found in *The Poetry Teacher's Book of Lists* as well as on Sylvia Vardell's Poetry for Children blog. Most of the established poetry blogs participate in the "Poetry Friday" celebration, posting a poem or poetry-related items on Fridays. Some include teaching activities and even welcome child participation. Sites and blogs also offer links to additional poetry resources on the web. See our select list of electronic resources that are particularly helpful in sharing poetry with children.

## 25 Children's Poetry Websites and Blogs You Need to Know

*About Poetry:*

**Alphabet Soup**
by Jama Rattigan
JamaRattigan.com

**The Academy of American Poets**
Poets.org

**Giggle Poetry**
GigglePoetry.com

**The Miss Rumphius Effect**
by Tricia Stohr-Hunt
MissRumphiusEffect.Blogspot.com

**Poetry Alive**
PoetryAlive.com

**Poetry for Children**
by Sylvia Vardell
PoetryForChildren.Blogspot.com

**Poetry Foundation**
PoetryFoundation.org

**The Poetry Minute**
PoetryMinute.org

**Poetry At Play: Poetry Advocates for Children and Young Adults**
PoetryatPlay.org

**Potato Hill Poetry**
Potatohill.com

**Wordswimmer** by Bruce Black
Wordswimmer.Blogspot.com

**A Year of Reading**
by Franki Sibberson
and Mary Lee Hahn
ReadingYear.Blogspot.com

*Poets:*

**April Halprin Wayland**
TeachingAuthors.com

**David L. Harrison's** blog
DavidLHarrison.Wordpress.com

**The Drift Record**
by Julie Larios
JulieLarios.Blogspot.com

**Father Goose**
by Charles Ghigna
CharlesGhigna.Blogspot.com

**Florian Café**
by Douglas Florian
FlorianCafe.Blogspot.com

**GottaBook**
by Greg Pincus
Gottabook.Blogspot.com

**Nikki Sounds Off**
by Nikki Grimes
NikkiGrimes.com/blog

**The Poem Farm**
by Amy Ludwig VanDerwater
PoemFarm.AmyLV.com

**Poetry for Kids**
by Kenn Nesbitt
Poetry4Kids.com

**Poetry Suitcase**
by Janet Wong
PoetrySuitcase.com

**Writing the World for Kids**
by Laura Purdie Salas
LauraSalas.com/blog

# Fun Websites to Support Science Learning

*In the* Take 5! *activities that accompany each poem in this book, we often reference websites that provide helpful information, visuals, diagrams, and more to support student learning in science. Here are some of those fun, interesting, and instructional resources.*

| | |
|---|---|
| AllAboutBirds.org | Kids.NationalGeographic.com |
| Animals.NationalGeographic.com | KidsButterfly.org |
| AnimalSpot.net | KidsHealth.org |
| ArborDay.org | Mission-blue.org |
| AustralianMuseum.net.au | NationalGeographic.com |
| BrainPop.com | NASA.gov |
| Burpee.com | NobelPrize.org |
| Calm.com | Optics4kids.org |
| Census.gov | OrganicGardening.com |
| Chem4kids.com | Panda.org |
| CicadaMania.com | Papertoys.com |
| CloudAppreciationSociety.org | Photography.NationalGeographic.com |
| CompostingCouncil.org | PhotosynthesisForKids.com |
| ConserveTurtles.org | RachelCarson.org |
| DiscoverE.org | RocksForKids.com |
| DrawingsofLeonardo.org | ScienceBuddies.org |
| EarthPopUpBook.weebly.com | SmithsonianEducation.org/Scientist |
| EyeChartMaker.com | SolarEnergy.org |
| FamousScientists.org | SoundCloud.com |
| FreeSound.org | Space.com |
| FrisbeeDisc.com | SpaceCamp.com |
| GalaxyMap.org | TeacherVision.fen.com |
| GetCaughtEngineering.com | TheReptileReport.com |
| Glogster.com | Tree-Pictures.com |
| HealthyPet.com | TryScience.org |
| HeavyEquipment.com | Tsunami.org |
| HistoryofBridges.com | Video.NationalGeographic.com |
| Howcast.com | Water.org |
| HowStuffWorks.com | Weather.com |
| Illusions.org | WebElements.com |
| IntoTheWind.com | Windustry.org |
| Inventions.org | WorldFishingNetwork.com |
| JaneGoodall.org | |

**Note:** Of course, we can't guarantee that these sites will remain active indefinitely or vouch for their ongoing accuracy. Please use your own judgment to assess their appropriateness for your audience.

# Websites and Blogs for Science Teaching K-5

*As we connect poetry and science, it can be helpful to know about resources that provide background for building science understanding. This select list focuses on science teaching with children and includes blogs, websites, videos, and more.*

**National Science Teachers Association**
NSTA.org
SciLinks.org
NSTACommunities.org/blog
NSTA.org/publications/freebies.aspx

**Seeds of Science/Roots of Reading**
ScienceandLiteracy.org

**Lawrence Hall of Science**
**University of California, Berkeley**
LawrenceHallofScience.org

**Understanding Science**
UndSci.Berkeley.edu

**Bill Nye, The Science Guy**
BillNye.com

**Author Seymour Simon's Blog**
SeymourSimon.com/index.php/blog/

**Star Walk Kids Media**
StarWalkKids.com

**Edheads Science and Math Games**
EdHeads.org

**Mr. Parr's YouTube Videos**
YouTube.com/user/ParrMr

**SciShow Science Videos on YouTube**
YouTube.com/scishow

**Bookish Ways in Math and Science**
BookishWays.Blogspot.com

**Science for Kids**
ScienceForKidsblog.Blogspot.com

**Growing a STEM Classroom**
GrowingaSTEMclassroom.Blogspot.com

**Mrs. Harris Teaches Science**
MrsHarrisTeaches.com

**Science Teaching Junkie**
TeachingJunkie.Blogspot.com

**The Simply Scientific Classroom**
MrsJacobsClassroom.Blogspot.com

**The Science Vault**
TheScienceVault.net

**Little Miss Hypothesis**
LittleMissHypothesis.Blogspot.com

**Science for All**
TeachScience4All.Wordpress.com

# Professional Resources

*This abbreviated list of professional reference sources will provide additional background that you will find helpful in selecting and sharing science poetry with young people. For further reading, you will find several dozen professional resources listed in* The Poetry Teacher's Book of Lists.

Akerson, Valarie. Teaching Science When Your Principal Says "Teach Language Arts" in *Teaching Teachers: Bringing First-Rate Science to the Elementary Classroom.*

Bauer, Caroline Feller. *The Poetry Break: An Annotated Anthology with Ideas for Introducing Children to Poetry.*

Booth, David and Moore, Bill. *Poems Please! Sharing Poetry with Children.*

Chatton, Barbara. *Using Poetry Across the Curriculum.*

Collom, Jack and Noethe, Sheryl. *Poetry Everywhere: Teaching Poetry Writing in School and in the Community.*

Cullinan, Bernice E., Scala, Marilyn C., and Schroder, Virginia C. *Three Voices: An Invitation to Poetry Across the Curriculum.*

Fitch, Sheree and Swartz, Larry. *The Poetry Experience: Choosing and Using Poetry in the Classroom.*

Franco, Betsy. *Conversations with a Poet: Inviting Poetry into K-12 Classrooms.*

Heard, Georgia. *Awakening the Heart: Exploring Poetry in Elementary and Middle School.*

Holbrook, Sara and Salinger, Michael. *Outspoken!: How to Improve Writing and Speaking Through Poetry Performance.*

Holbrook, Sara. *Practical Poetry: A Nonstandard Approach to Meeting Content-Area Standards.*

Honey, Margaret and Kanter, David E. *Design, Make, Play: Growing the Next Generation of STEM Innovators.*

Kennedy, X. J. and Kennedy, Dorothy. *Knock at a Star.*

Morgan, Emily and Ansberry, Karen. *Even More Picture-Perfect Science Lessons, K-5: Using Children's Books to Guide Inquiry.*

Partington, Richie. *I Second that Emotion: Sharing Children's and Young Adult Poetry: A 21st-Century Guide for Teachers and Librarians.*

Royce, Christine Anne, Morgan, Emily, and Ansberry, Karen. *Teaching Science through Trade Books.*

Sloan, Glenna. *Give Them Poetry: A Guide for Sharing Poetry with Children K-8.*

Sousa, David A. and Pilecki, Tom. *From STEM to STEAM: Using Brain-Compatible Strategies to Integrate the Arts.*

Vardell, Sylvia. *Poetry Aloud Here 2: Sharing Poetry with Children.*

Vardell, Sylvia. *Poetry People: A Practical Guide to Children's Poets.*

Vardell, Sylvia. *The Poetry Teacher's Book of Lists.*

# E-Resources for Poetry Teaching

One of the most controversial topics in the world of reading today concerns e-books. Some people think that e-books will replace paper books and change the way we read—and they're afraid of those changes. We agree that changes will happen, but we're excited by the possibilities and we offer both print and e-book versions of this book. Consider:

- a teacher can read a book review at lunch and buy an e-book version of the book (for less than the price of lunch);
- that book might be a collection of poems from Mexico or Australia but is delivered immediately without shipping costs or customs fees;
- the teacher can download the e-book onto an e-reader or a regular computer and project it onto a screen for the whole class to read aloud together;
- e-resources are easily searchable. A teacher can look for a poem using keywords like *solar system* or *armadillo*. Even if you prefer paper books, you might consider owning a second copy that is digital as a teaching resource;
- and reluctant readers (who might not like paper books but might enjoy manipulating text on a screen) can read the book using electronic bookmarks, a glossary, and sometimes read-aloud features, too.

Poetry is particularly well-suited to e-books. Imagine: a second grader is standing in line at the post office with his mother. He is bored. His mother hands him her cell phone. To play a video game? No—to read a poem in an e-book. He reads the short poem to himself and likes it. Then he reads it again—as he's been taught—aloud. His mother laughs. The woman standing behind him laughs. He reads another poem aloud and directs his mother (in a second reading) to chime in and guess the rhyming word. The man in front of him turns around to say the rhyming words. Next thing they know, the boy and his mother are first in line and enjoying reading together.

If you want to try an e-book and need to know where to get started, you'll find some titles and resources here:

PoetryForScience.Blogspot.com

PoetryFridayAnthology.Blogspot.com

PFAMS.Blogspot.com

PoetryTagTime.Blogspot.com

TeenPoetryTagTime.Blogspot.com

PoetryGiftTag.Blogspot.com

PoetryforChildren.Blogspot.com

PoetryTeachersBookofLists.Blogspot.com

# A Mini-Glossary of Science Terms

*Science poems incorporate vocabulary that can be challenging for students. Here is a short list of key terms and definitions used throughout the poems.*

**Acceleration**: To speed up and go faster and faster

**Aerodynamics**: The design of objects to allow them to move easily through air with reduced friction

**Animation**: The motion of characters on video created by a series of slightly changing individual images

**Atom**: The smallest bit of matter that still holds its own chemical properties

**Avian:** A term referring to birds

**Bacteria**: Single-celled microscopic life forms having no nuclear membrane holding genetic material

**Beaker**: A glass container with a straight side for measuring volume in laboratories

**Biodiversity**: The wide variety of all life forms found on the earth

**Bioprinting**: Creating new, living tissue from cells in a 3-D process in labs

**Budding**: The small bumps on a tree that grow into new leaves, shoots, or flowers

**Calories**: A measure of the amount of energy held in food

**Camouflage**: Natural protective coloration on living organisms that helps them blend in with their surroundings

**Capillary action**: The upward force of a liquid due to water molecules' tendency to stick together in tight spaces

**Carbs** (carbohydrates): The nutrients in food that supply energy; also a short name for *carbohydrates*

**Casts:** The kind of fossil formed when a mold is filled

**Celsius**: The scale for measuring temperature based on 100, with 0 for freezing and 100 for boiling

**Chlorophyll**: The green substance in plants that captures sunlight to harness its energy to make food

**Cicada**: A group of insects having a life cycle that varies in the number of years it appears, noticeable for the loud buzzing by the males

**Cinder cone**: The most common kind of volcano, and one that forms around a central vent or column in which the lava flows down the sides

**Compost**: A nutrient-rich material formed from decaying organic matter like leaves, vegetable parts, and manure

**Cretaceous:** A geologic period of time dating from 140 million to 65 million years ago during the time dinosaurs developed and then became extinct

**Deforestation**: The removal of all trees in a particular region or area

**Drag**: The force caused by the resistance of the air to an object moving forward

**Drought**: A period of time with little or no rain

**Eco-agricultural**: An integrated ecosystem approach to agriculture that will sustain rural farming, conserve biodiversity, and develop sustainable farms

**Electromagnetic spectrum**: The range or scale of radiation from the sun in decreasing wavelength size, such as radio, microwaves, visible light, and x-rays

**Element**: Matter that cannot be broken down into any other substance and still keeps its same properties

**Elliptic**: The shape of a flattened oval

**Engineer**: A person who is specially trained to design and build machines or buildings and create solutions to problems

**Evaporation**: The process in which molecules in a liquid move into the air as a result of heat speeding up their motion

**Exposure:** Film coming into contact with light to produce a picture

**Flask**: A specially shaped container with a narrow neck used in science labs

**Fulcrum**: The point on which a lever rests and turns about

**Fungi**: The plant-like organisms that take nourishment by growing on other plants or decaying material; singular is *fungus*

**Gall bladder**: A pouch-like body organ under the liver that secretes bile to help digest fats

**Gear**: A toothed wheel that uses a change of direction to turn other toothed wheels to determine the rate of motion

**Genes**: The hereditary material in the nucleus of cells that determines characteristics for developing cells

**Genetic**: Having inherited traits or characteristics from one's parents

**Geologist**: A scientist who studies the earth and its processes

**Graft**: To insert a cutting from one plant into another plant so that they grow together as one plant

**Graph**: A chart or diagram that shows the relationship of one thing to another in numbers or amounts

**Gravity**: The attractive force between two objects that pulls the smaller one toward the larger one

**Greenhouse gases**: Carbon dioxide and other gases that trap extra heat from the sun and warm the earth's atmosphere

**$H_2O$**: The chemical symbol for water signifying two atoms of hydrogen bonded with one atom of oxygen to form water

**Humidity**: A high amount of water vapor in the air at a given time

**Humus**: Rich soil formed by the decay of plants and animals

**Hybrid**: An automobile that runs on both gasoline and electrical power from a battery

**Hypothesis**: The knowledgeable prediction an experimenter states in order to design an experiment to learn if the answer to a question is accurate

**Immunity**: The body's ability to build up defensive cells that prevent illnesses caused by germs

**Indigo**: A dark purplish-blue color on the visible spectrum

**Lever**: A simple machine made of a bar and a fulcrum around which the bar pivots to lift loads

**Lift**: The force that moves an airplane upward caused by the forward motion of the plane and its wings

**Molecule**: The smallest bit of matter from a compound (made of two or more elements) that still retains the properties of that matter

**Natural selection**: The concept that all species evolved from common ancestors and that the best and most fit survived to pass along their characteristics while the weaker ones died out

**Naturalist**: A person who studies nature and its science

**Near Earth Object (NEO)**: The comets and asteroids affected by the gravity of other planets that enter into orbit around the earth

**Nebula**: The cloud of dust and gas visible in space as a bright area at night

**Niche**: A place in the environment and an organism's role there

**Nymph**: The young form of some insects in which the young resemble the parents and shed their outer shell multiple times to fit their growing body

**Observation**: The information gained directly using the senses or exact measurements

**Organism**: A living thing like a plant or animal

**Outsource:** To get products from a foreign or outside supplier

**Pachycephalic:** A kind of skull where the bone is extra thick

**Paleoentologist:** A scientist who studies life that lived during ancient or geological times

**Pancreas**: A body organ that regulates the amount of sugar in the blood

**Photons**: Tiny particles that make up light waves

**Photosynthesis**: The food-making process carried out by green plants using the sun's energy, water, and carbon dioxide

**Plankton**: Microscopic, single-celled life in the water

**Pressure**: The continuous physical force applied on or to an object

**Prism**: A clear solid object with a triangular base and flat sides that breaks visible light into the colors of the spectrum.

**Radiant energy**: Energy traveling in the form of electromagnet waves, including radio waves, visible light, and x-rays

**Ratio**: A numerical relationship between two numbers in a specific proportion

**Reservoir**: A reserve of water in a lake or collecting pool stored for later use

**Rotor**: The part of a machine that rotates or spins

**Saturation:** The intensity of color

**Seacow**: The informal name for a manatee

**Sirenian**: Having to do with manatees and dugongs

**Slash-and-burn**: A way of adding more land for farming by cutting and burning the plants that grow there naturally; often done in forests

**Solar panel**: A flat structure lined with solar cells that absorb and collect the sun's energy which is then converted into electrical energy

**Sphere**: A round, solid figure with every surface point an equal distance from the midpoint

**Statistics**: The practice of collecting, organizing, and making sense of number data

**Symmetry**: A pattern in which exactly alike parts face opposite one another along a plane or center division or arranged equally around a midpoint

**Traits**: The qualities or characteristics of a living organism

**Tuberculosis**: A very contagious lung disease caused by bacteria

**Turbine**: A machine for generating power with a bladed wheel caused to rotate by water, wind, or gas passing over the blades

**Vacuum**: A sealed container holding no air or gases

**Vapor**: A gas formed after evaporation

**Variable**: The factor in an experiment that is controlled or changed

**Vertex**: The meeting point of two lines

**Vibrate**: To move or shake back and forth quickly

**Voltage**: The measure of the force of an electrical current

**Voltmeter**: An instrument for measuring the difference in potential between varying points of an electrical circuit in units called *volts*

**Watershed:** An area of land where water empties into rivers and lakes

# Indexes and Credits

# Title Index

**A**
Accidentally On Purpose 233
After I Made a Huge Mess 159
Albert Einstein 219
Alligator with Fish 49
Armor 156
Astronauta común, Un 104
At the Speed of Light 240
Auntie V's Hybrid Car 48
Ay agua, mi amiga 204

**B**
Backwards 71
Becoming Butterflies 131
Besado por el sol 171
Big Sun 43
Biological Community, A 192
"Black Leonardo," The 100
Breakfast Alchemy 174
Brink, The 158

**C**
Camouflage 170
Cancer 183
Can Our Eyes Fool Our Taste Buds? 54
Capillary Action 34
Celsius Thermometer 76
Changes 197
Chicharra 251
Cicada 251
Cicada Magic 212
Citizen Scientist 172
Class Plant, The 113
Classroom in the Meadow 152
Climate Versus Weather 245
Clouds 85
Comet Hunter 242
Compu-nerdo 61
Computer Geek 61
Computer Models 196
Comunidad biológica, Una 192
Considering Copernicus 179
Cool Food for Thought 249
Crane Operator, The 177
Crazy Data Day 115

**D**
Da Vinci Did It! 99
Dear Rachel Carson 89
Descubrimiento 112
Designing an Experiment 234
Did You Know? 42
Dinos in the Laboratory 190
Discovery (Engle) 112
Discovery (Kennedy) 73
Dog in a Storm 45
Dog's Hypothesis, A 193
Dr. Lee 62
Driftwood Hut 97

**E**
Earth's Tilt 123
Elemental 238
Engineer, The 101
Everyday Astronaut 104

**F**
First Science Project 94
Five O'Clock Rush 176
Food for Thought 214
Fossil Fuels 128
Foundation (Don't Rush It!) 217
Frames Per Second (fps) 255
Friction 200
Frisbee 80
Froggy 194
Future Dreams Idea #63 64

**G**
Galileo Galilei 180
Game Programmer 135
Gears 136
Geologist 102
Glacier 127
Go Fly a Kite 120
Going Bananas 235
Grafting 210
Gravity 119
Great Pyramid of Giza, The 257

## H
Hand-Me-Downs 72
Hands 53
Harbor Wave at Hilo: Tsunami Survivor 246
Hawking Time 220
Hello, Hello! 55
How to Be a Scientist 69
Hurricane Hideout 206

## I
I Have a Question 31
I Like that Night Follows Day 92
I Want to Know Why 103
I Will Be a Chemist: Mario José Molina 260
Imagine Small 117
Inherit Tense 52
Inquiry 151
Invention Intentions 182

## J
Jane Goodall Begins a Speech 139
Jóvenes y viejos juntos 51

## L
Lament of Lonesome George, The 143
Late Night Science Questions 111
Leopard Cannot Change His Spots, The 130
Let Me Join You 253
Let's All Be Scientists! 109
Levers 96
Life Cycle 84
Lift 160
Lion and the House Cat, The 90
Liquids Can't Contain Themselves 118
Listen 41
Looking at the Sky Tonight 82
Love Note to a Magnet 79
Lunar Eclipse 163

## M
Magic Show 87
Meet Mr. Wizard 154
Metal Monster 56
Meter Stick 116
Microwave Oven 216
Mold 134
Moving for Five Minutes Straight 213
Moving to Atlantis City, 2112 184
My Bean Plant 35
My Experiment 114
My Photo Experiment 221
My Project for the Science Fair 98
My Robot 137
My Rock 47
My WristRobot Pack 264

## N
NEO Hunters, The 202
New Dinosaur, A 262
No Penguins Here 239
Nursing Math 195

## O
Ocean Engine 244
Ocean Explorer Sylvia Earle 60
Oh Water, My Friend 204
Old Water 44
Orion Nebula 162
Our Truck 77

## P
Paper Airplanes 231
Pass Me Those Ear Muffs 110
Patterns in Nature 252
Photosynthesis 91
Pieces 95
Plates 126
Playground Physics 232
Printing, Pressed Beyond Words 261
Prisas a las cinco 176
Prism 81
Protecting My Friend 173
Pumpkin Experiment 74
Push Power 39

## Q
Queen of Night 203
Questions That Matter 157
Questions, Questions 191

## R
Rachel Carson 59
Rain Forest, The 129
Rain Gauge 125

Real Thing, The 224
Recycling 88
Resources Rule! 248
Riddle for a Dry Day 46
Riddle for a Wet Day 86
Rings Not Letters 132
Rocky Rescue 222
Roller Coaster Ride 199

## S
Science 265
Science Fair 218
Science Fair Day 58
Science Fair Project 178
Science Lab Pledge, The 70
Science Project 138
Scientific Inquiry 229
Scientific Steps 23
Seeing School 93
Shade-Grown 223
Shadow Grows, The 243
Shen Kuo 259
Shots! Shots! Shots! 133
Sink or Float 33
Snake Traits 50
Soda Machine Bite 256
Soil Inventory 247
Solar Power 208
Sound Waves 161
Sound Waves at Breakfast 121
Space Yacht 144
Step Outside. What Do You See? 32
Stopwatch 36
Sugar Water 78
Sun-Kissed 171
Superhero Scientist 30
(Super)Power: (to the) Point 141

## T
Take Backs 38
Teacher's Look 241
Tell It to the Court 258
Testing My Hypothesis 153
Testing My Magnet 75
Thank You, Isaac Newton 40
That Dish Thing 236

Things to Do in Science Class 150
Think of an Atom 237
Thirsty Measures 254
This Week's Weather 165
Tide Pool 169
Tinker Time 57
Titan in Man's Seaweed 263
To the Eye 201
Tornado! 166
Trilobite 167
Tropical Rain Forest Sky Ponds 209

## U
Uh Oh, Pluto 122

## V
Virtual Adventure 215
Voy a ser químico: Mario José Molina 260

## W
Water 63
Water Engineered 142
Water + Dirt = 37
Water Round 124
We Need Green Seaweed! 164
Weather Map 205
Welcome to the Science Lab 230
What Am I? 175
What Can You Make from Carbon? 198
What I Know about the Sun 83
What Is a Foot? 250
What Is Science? 189
What Makes a Turbine Turn 168
What We Eat 207
When You Are a Scientist 29
Which Ones Will Float? 149
Wiki Alert 181
Windfall in the Andrews Forest 211
Wondering Why 140

## Y
Young and Old Together 51

## Z
Zapped! 155

# Poet Index

**A**
Acey, Joy 34, 133
Ada, Alma Flor 260
Ashman, Linda 50
Atkins, Jeannine 122, 131, 152, 195, 232, 238

**B**
Bernier-Grand, Carmen T. 61
Black, Robyn Hood 214, 222, 261
Blackaby, Susan 88, 208, 229, 248
Brown, Susan Taylor 193
Bruchac, Joseph 207, 211
Bulion, Leslie 60, 124, 153, 244

**C**
Calmenson, Stephanie 45, 101
Campoy, F. Isabel 176
Carter, James 220, 265
Coombs, Kate 53, 63, 85, 93, 127, 247
Cotten, Cynthia 23, 151, 189

**D**
Dempsey, Kristy 141, 182, 190
Denton, Graham 110
Dotlich, Rebecca Kai 37, 177
Duke, Shirley Smith 140, 172, 240, 241, 252

**E**
Engle, Margarita 51, 112, 156, 164, 170, 192, 209, 223

**F**
Florian, Douglas 43, 123
Franco, Betsy 102, 213

**G**
Gerber, Carole 166
Ghigna, Charles 52, 84
Graham, Joan Bransfield 30, 205, 245

**H**
Hahn, Mary Lee 74, 89, 90, 159, 162, 183
Harley, Avis 234
Harrison, David L. 103, 137
Harshman, Terry Webb 203
Havill, Juanita 87, 132, 144, 202
Hershenhorn, Esther 175
Hoberman, Mary Ann 167
Holbrook, Sara 142, 200, 249
Hubbell, Patricia 79, 199

**J**
Jules, Jacqueline 173, 256

**K**
Katz, Bobbi 129, 163, 179
Kennedy, X.J. 56, 73

**L**
Larios, Julie 42, 59, 75, 114, 219
Latham, Irene 46, 86, 218
LaTulippe, Renée M. 76, 95, 97, 99, 109, 180, 215
Levy, Debbie 181
Lewis, J. Patrick 100
Lyon, George Ella 72, 154

**M**
McCall, Guadalupe Garcia 171, 204, 251
Mordhorst, Heidi 212

**N**
Nelson, Marilyn 262
Nesbitt, Kenn 98
Newman, Lesléa 94, 130

**O**
Ode, Eric 29, 58, 149, 178

**P**
Park, Linda Sue 224, 233
Paul, Ann Whitford 126, 191
Pincus, Greg 111

**Q**
Quattlebaum, Mary 174

**R**
Roemer, Heidi Bee 118, 157, 230, 235, 253, 254
Rosen, Michael J. 263
Ruddell, Deborah 70

**S**
Salas, Laura Purdie 120, 150, 198, 201, 243, 257
Salinger, Michael 96, 136, 161, 239
Schroeder, Glenn 80
Sidman, Joyce 119
Silverman, Buffy 237
Singer, Marilyn 91, 160
Slesarik, Ken 47
Spinelli, Eileen 40, 83, 117
Suen, Anastasia 31, 125
Swanson, Susan Marie 121, 139

**T**
Tafolla, Carmen 61, 104, 264
Thompson, Holly 242

**V**
VanDerwater, Amy Ludwig 35, 41, 69, 81, 116

**W**
Wardlaw, Lee 138
Waters, Charles 134, 194, 217
Wayland, April Halprin 44, 54, 92, 155
Weatherford, Carole Boston 246
Withrow, Steven 168, 184
Wolf, Allan 32
Wolff, Virginia Euwer 236
Wong, Janet 33, 36, 38, 39, 48, 55, 57, 62, 64, 71, 77, 82, 113, 115, 128

**Y**
Yolen, Jane 49, 143, 169, 250

# Subject Index

## A
acceleration 199
air 41, 45, 64, 77, 80, 91, 113, 119, 120, 152, 154, 161, 166, 184, 209, 241, 242, 244, 249, 260
alligator 49
anaconda 50
animated 61
animation 255
asteroid 202
astronaut 104
astronomers 122, 123, 259
atom 117, 237
avocado 94
axis 123

## B
bacteria 236
beaker 230, 238
Big Dipper 82
binoculars 242
biodiversity 244
bioprinting 261
biosphere 248
birds 59, 112, 129, 179, 215, 247
bolts 57
building 196, 217, 246
butterflies 131

## C
calculate 135, 195, 196
calories 214
canopy 129, 211, 248
capacity 254
capillary action 34
carbohydrates 214
carbon 164, 198, 249
carbon emissions 249
Carson, Rachel 59, 89
Carver, George Washington 100
caterpillar 131
caverns 248
cells 183, 189, 208, 236, 261
Celsius 76
chemical change 197
chemicals 89, 230
chimpanzee 139

chlorophyll 249
chrysalis 131
cicadas 212, 251
cinder cone 102
climate 244, 245, 259
cloning 236
clouds 45, 85, 119, 124, 144, 205, 237, 244
cobra 50
colors 81
comet 122, 242
community 192
compass 259
compost 87
computer 57, 61, 101, 135, 196
conclusion 23, 222
cone 102, 138
constellation 242
copy 258
cranes 257
cup 35, 38, 63, 82, 158, 174, 254, 264
cures 103, 183
Curie, Madame Marie 238
cyclone 206

## D
Darwin, Charles 140
data 23, 115, 149, 172, 229, 235
DaVinci, Leonardo 99, 100
day 42, 46, 48, 58, 83, 86, 92, 94, 104, 113, 115, 116, 129, 177, 203, 205, 242, 245, 246, 264
deforestation 223
dinosaur 262
diseases 86, 103
dissect 194
doctor 62, 113
Douglas fir 211
drag 120
dreams 64, 203, 217
driftwood 97
drought 46

## E
Earle, Sylvia 60

Earth 59, 60, 83, 119, 123, 144, 189, 202, 204, 208, 210, 240, 242, 244, 246, 247, 248, 260, 265
earthquake 64, 126
Einstein, Albert 219
elbow 253
electricity 73
electromagnetic radiation 216
electron 156, 208
element 238, 260
energy 161, 208, 232, 240
engineer 99, 101, 142, 196
evaporate 124, 204, 244
experiment 23, 69, 74, 75, 109, 114, 137, 153, 221, 234
extinct 190, 222
eyes 43, 44, 52, 54, 69, 90, 92, 104, 117, 201, 202, 205, 206, 246

## F
family tree 52
feet 94, 102, 126, 150, 152, 153, 213, 243, 250
flasks 236
float 33
flood 86
food chain 249
food label 214
force 96, 168, 232

## G
game designer 61
gear 30, 57
gears 57, 136, 208
genes 72
genetic 72
genetics 218
geologist 102, 259
glasses 62, 93, 190
goggled 70
goggles 30, 230
Goodall, Jane 139
grafts 210
graph 35
gravity 40, 119, 160, 232
greenhouse gases 244

## H
$H_2O$ 44
hands 30, 39, 53, 152, 203, 211, 218, 230, 241
Hawking, Steven 220
heart 60, 119, 194, 205, 213, 265
humus 248
hurricane 206
Hyakutake, Yuji 242
hybrid 48
hypothesis 23, 153, 193, 229
hypothesize 69

## I
ice 122, 127, 158, 242, 244
ingredients 214
inquiry 151, 229
invention 101, 160, 182
inventor 100
investigation 109
invisibility 224

## J
joints 141, 253

## K
kite 73, 119, 120, 144, 168
Kuo, Shen 259

## L
lab 70, 74, 154, 184, 230, 236, 260
lava 102
leaves 34, 48, 50, 87, 94, 113, 152, 171, 234, 247
lever 96, 177, 232, 257
lichen 211
lift 120, 136, 160, 177, 232
light 81, 83, 85, 91, 92, 94, 113, 144, 159, 189, 201, 203, 206, 208, 219, 240
lightning 45, 73, 166, 240
limestone 257
lion 90, 246, 249
liquids 118, 157, 174
living 47, 55, 56, 63, 89, 143, 169, 198, 245

**M**
machine(s) 56, 96, 98, 110, 136, 175, 208, 256, 257
magnet 75, 79, 239
magnetic 239, 259
manatees 263
measure(s) 76, 116, 151, 152, 155, 174, 195, 230, 242, 254
Mendel, Gregor 218
meteorologist 205
microscope 30, 136, 156, 260
microwaves 176, 216, 240
Milky Way 180, 203
minerals 214, 248
models 196, 224, 231
mold(y) 113, 134, 261
molecule(s) 117, 216, 260
Molina, Mario José 260
moon 92, 104, 119, 163, 180, 203
motion 168, 189, 208, 232
mud 37, 39, 83, 86, 128, 137, 212

**N**
natural selection 140
nature 89, 238, 252, 263, 265
Near Earth Object (NEO) 202
niche 182, 192, 209
night 85, 92, 104, 111, 125, 129, 135, 153, 165, 168, 177, 179, 203, 222, 242, 247
north 168, 239, 259, 262
nutrients 244
nuts 57

**O**
observation(s) 23, 151, 189, 229
ocean 60, 111, 124, 164, 168, 184, 189, 237, 244, 246, 263
orbit(s) 122, 123, 242
ore 248
organic 89
organism(s) 192, 249
outside 32, 237

**P**
patterns 205, 252
peanut 100

Perey, Marguerite 238
Petri dish 236
Petri, Richard Julius 236
phone 55, 191
photons 144
photosynthesis 91
physical change 197
physics 232, 264
planet 60, 89, 104
plankton 128, 244
plant(s) 35, 53, 74, 83, 87, 91, 94, 113, 116, 128, 164, 209, 223, 233, 249
planting 51
Pluto 122
prediction 113
pressure 120, 206
prism 81
protein 214, 233
pulleys 257
pulse 213
pumpkin 74, 117
push(es) 39, 45, 53, 144, 159, 163, 213, 239
pyramid 257

**Q**
question(s) 23, 31, 52, 109, 111, 140, 151, 157, 182, 189, 191, 231

**R**
radiant energy 240
radish 51
rain 32, 45, 46, 84, 125, 132, 137, 152, 205, 209, 211, 245
raining 63
rain forest 129, 209, 223
rain gauge 125
raindrops 124
ratio 136
recycling 88, 164
report(ed) 242, 258
resources 208, 248
results 23, 75, 189, 193, 229
rivers 49, 63, 84, 165, 248, 265
robot 56, 101, 111, 137
robotics 264

rocks 38, 47, 102, 119, 122, 127, 169, 247
roots 51, 94, 209, 211, 247
rust(y) 75, 77

**S**
safety 30, 190, 230
salt 38, 97, 214
sand 38, 102, 203, 232, 246
science 58, 70, 94, 98, 110, 111, 138, 150, 154, 155, 159, 174, 189, 190, 194, 218, 221, 222, 230, 262, 264, 265
science fair 58, 178, 218, 221
scientist(s) 29, 30, 42, 69, 100, 139, 172, 229, 232, 234, 259
seasons 123
seeds 51, 74, 247
serving size 214
shadow 163, 243
shots 133
sick 103, 113, 134, 173, 195
sink 33, 102, 149
snake 50, 75
soil 100, 113, 124, 207, 209, 211, 233, 247, 249
solar cells 208
solar panel 249
solar power 208
sound 41, 59, 110, 121, 137, 161, 189, 200, 213, 241, 265
space 44, 82, 104, 119, 144, 184, 191, 202, 208, 209, 237, 240, 265
spring 59, 89, 173, 248
stars 43, 82, 92, 130, 162, 180, 242, 257, 265
statistics 195
stems 51, 247
stopwatch 36, 115
storm 45, 165, 166, 206, 244
storm surge 206
stream 84, 211
submarine 60
sun 43, 46, 71, 83, 94, 122, 132, 169, 171, 179, 180, 184, 197, 201, 203, 204, 205, 207, 208, 212, 240, 243, 244, 249, 251, 257
symmetry 252

**T**
taste 54, 152, 193, 210
thermometer 76
thunder 45, 85 tides 97, 169, 203, 246
time(s) 196, 203, 210, 220, 221, 224
time machine 98
tools 110, 159, 230, 257
tortoises 140
traits 50, 140
tree(s) 32, 40, 48, 52, 84, 94, 95, 119, 121, 129, 132, 177, 209, 210, 211, 249, 251
tropical rain forest 209
tsunami 246
turbines 168, 208
TV 95

**U**
understory 129

**V**
vaccinations 133
valley 127
variable 234
Venus 42, 180
vibrate 41, 216
vibrations 121, 241
video(s) 53, 55, 95, 215, 255
video games 53, 135, 255
virtual 215, 261
vitamins 214
volcano 102

**W**
water 37, 44, 59, 63, 76, 77, 78, 87, 91, 94, 113, 124, 125, 127, 128, 142, 161, 204, 206, 216, 240, 246, 248, 260
watershed 248
wave(s) 41, 86, 161, 121, 144, 240, 241, 246
weather 45, 113, 165, 172, 205, 206, 245
wells 63
wheels 58, 101, 200

# Copyright & Permissions

For permission to reprint any of the poems in this book, please contact the individual poets listed here either directly or through their agents.

Most of these poets can be reached through their individual websites, which are listed at our Pomelo Books website, PomeloBooks.com. If you need help getting in touch with a poet, just let us know and we'll be happy to connect you.

A note on copyright:

**If it doesn't feel right to copy it . . . please don't!**

Poets (like plumbers and lawyers and teachers and acrobats) need to earn a living from their work; permissions fees and royalties help pay the rent!

# Poem Credits

**Joy Acey:** "Capillary Action" (Kindergarten, Week 6: Investigations), "Shots! Shots! Shots!" (2nd Grade, Week 25: Human Body); copyright ©2014 by Joy Acey. Used with permission of the author. All rights reserved.

**Alma Flor Ada**: "I Will Be a Chemist: Mario José Molina / Voy a ser químico: Mario Molina" (5th Grade, Week 32: More Famous Scientists); copyright ©2014 by Alma Flor Ada. Used with permission of the author. All rights reserved.

**Linda Ashman**: "Snake Traits" (Kindergarten, Week 22: Adaptations & Traits); copyright ©2014 by Linda Ashman. Used with permission of the author. All rights reserved.

**Jeannine Atkins:** "Uh Oh, Pluto" (2nd Grade, Week 14: Space), "Becoming Butterflies" (2nd Grade, Week 23: Cycles), "Classroom in the Meadow" (3rd Grade, Week 4: Observations), "Nursing Math" (4th Grade, Week 7: Data), "Playground Physics" (5th Grade, Week 4: Observations), "Elemental" (5th Grade, Week 10: More Matter); copyright ©2014 by Jeannine Atkins. Used with permission of the author. All rights reserved.

**Carmen T. Bernier-Grand:** "Computer Geek/Compu-nerdo" (Kindergarten, Week 33: Computers); copyright ©2014 by Carmen T. Bernier-Grand. Used with permission of the author. All rights reserved.

**Robyn Hood Black:** "Food for Thought" (4th Grade, Week 26: Kitchen Science), "Rocky Rescue" (4th Grade, Week 34: Science Careers), "Printing, Pressed Beyond Words . . ." (5th Grade, Week 33: Computers); copyright ©2014 by Robyn Hood Black. Used with permission of the author. All rights reserved.

**Susan Blackaby:** "Recycling" (1st Grade, Week 20: Natural Resources), "Solar Power" (4th Grade, Week 20: Natural Resources), "Scientific Inquiry" (5th Grade, Week 1: Scientific Practices), "Resources Rule!" (5th Grade, Week 20: Natural Resources); copyright ©2014 by Susan Blackaby. Used with permission of the author. All rights reserved.

**Susan Taylor Brown:** "A Dog's Hypothesis: Zoey's Guide to Getting More Goodies" (4th Grade, Week 5: Predictions & Hypotheses); copyright ©2014 by Susan Taylor Brown. Used with permission of the author. All rights reserved.

**Joseph Bruchac:** "What We Eat" (4th Grade, Week 19: Soil & Land), "Windfall in The Andrews Forest" (4th Grade, Week 23: Cycles); copyright ©2014 by Joseph Bruchac. Used with permission of the author. All rights reserved.

**Leslie Bulion:** "Ocean Explorer Sylvia Earle" (Kindergarten, Week 32: More Famous Scientists), "Water Round" (2nd Grade, Week 16: The Water Cycle), "Testing My Hypothesis" (3rd Grade, Week 5: Predictions & Hypotheses), "Ocean Engine" (5th Grade, Week 16: The Water Cycle); copyright ©2014 by Leslie Bulion. Used with permission of the author. All rights reserved.

**Stephanie Calmenson:** "Dog in a Storm" (Kindergarten, Week 17: Weather & Climate), "The Engineer" (1st Grade, Week 33: Computers); copyright ©2014 by Stephanie Calmenson. Used with permission of the author. All rights reserved.

**F. Isabel Campoy:** "Five O'Clock Rush" / "Prisas a las cinco" (3rd Grade, Week 28: Machines); copyright ©2014 by F. Isabel Campoy. Used with permission of the author. All rights reserved.

**James Carter:** "Hawking Time" (4th Grade, Week 32: More Famous Scientists), "Science" (A Poem for Everyone—last poem); copyright ©2014 by James Carter. Used with permission of the author. All rights reserved.

**Kate Coombs:** "Hands" (Kindergarten, Week 25: Human Body), "Water" (Kindergarten, Week 35: Future Challenges), "Clouds" (1st Grade, Week 17: Weather & Climate), "Seeing School" (1st Grade, Week 25: Human Body), "Glacier" (2nd Grade, Week 19: Soil & Land), "Soil Inventory" (5th Grade, Week 19: Soil & Land); copyright ©2014 by Kate Coombs. Used with permission of the author. All rights reserved.

**Cynthia Cotten:** "Scientific Steps" (A Poem for Everyone—first poem), "Inquiry" (3rd Grade, Week 3: Ask and Ask Again), "What Is Science?" (4th Grade, Week 1: Scientific Practices); copyright ©2014 by Cynthia Cotten. Used with permission of the author. All rights reserved.

**Kristy Dempsey:** "(Super)Power: (to the) Point" (2nd Grade, Week 33: Computers), "Invention Intentions" (3rd Grade, Week 34: Science Careers), "Dinos in the Laboratory" (4th Grade, Week 2: Lab Safety); copyright ©2014 by Kristy Dempsey. Used with permission of the author. All rights reserved.

**Graham Denton:** "Pass Me Those Ear Muffs" (2nd Grade, Week 2: Lab Safety); copyright ©2014 by Graham Denton. Used with permission of the author. All rights reserved.

**Rebecca Kai Dotlich:** "Water + Dirt =" (Kindergarten, Week 9: Matter), "The Crane Operator" (3rd Grade, Week 29: Building Things); copyright ©2014 by Rebecca Kai Dotlich. Used with permission from Curtis Brown, Ltd. All rights reserved.

**Shirley Smith Duke:** "Wondering Why" (2nd Grade, Week 32: More Famous Scientists), "Citizen Scientist" (3rd Grade, Week 24: Patterns), "At the Speed of Light" (5th Grade, Week 12: More Force, Motion & Energy), "Teacher's Look" (5th Grade, Week 13: Light & Sound), "Patterns in Nature" (5th Grade, Week 24: Patterns); copyright ©2014 by Shirley Smith Duke. Used with permission of the author. All rights reserved.

**Margarita Engle:** "Young & Old Together" / "Jovenes y viejos juntos" (Kindergarten, Week 23: Cycles), "Discovery" / "Descubrimiento" (2nd Grade, Week 4: Observations), "Armor" (3rd Grade, Week 8: Tools of Science), "We Need Green Seaweed!" (3rd Grade, Week 16: The Water Cycle), "Camouflage" (3rd Grade, Week 22: Adaptations & Traits), "A Biological Community" / "Una comunidad biologica" (4th Grade, Week 4: Observations), "Tropical Rain Forest Sky Ponds" (4th Grade, Week 21: Ecosystems), "Shade-Grown" (4th Grade, Week 35: Future Challenges); copyright ©2014 by Margarita Engle. Used with permission of the author. All rights reserved.

**Douglas Florian:** "Big Sun" (Kindergarten, Week 15: Sun, Earth & Moon), "Earth's Tilt" (2nd Grade, Week 15, Sun, Earth & Moon); copyright ©2014 by Douglas Florian. Used with permission of the author. All rights reserved.

**Betsy Franco:** "Geologist" (1st Grade, Week 34: Science Careers), "Moving for Five Minutes Straight" (4th Grade, Week 25: Human Body); copyright ©2014 by Betsy Franco. Used with permission of the author. All rights reserved.

**Carole Gerber:** "Tornado!" (3rd Grade, Week 18: Forces of Nature); copyright ©2014 by Carole Gerber. Used with permission of the author. All rights reserved.

**Charles Ghigna:** "Inherit Tense" (Kindergarten, Week 24: Patterns), "Life Cycle" (1st Grade, Week 16: The Water Cycle); copyright ©2014 by Charles Ghigna. Used with permission of the author. All rights reserved.

**Joan Bransfield Graham:** "Superhero Scientist" (Kindergarten, Week 2: Lab Safety), "Weather Map" (4th Grade, Week 17: Weather & Climate), "Climate Versus Weather" (5th Grade, Week 17: Weather & Climate); copyright ©2014 by Joan Bransfield Graham. Used with permission of the author. All rights reserved.

**Mary Lee Hahn:** "Pumpkin Experiment" (1st Grade, Week 6: Investigations), "Dear Rachel Carson" (1st Grade, Week 21: Ecosystems), "The Lion and the House Cat" (1st Grade, Week 22: Adaptations & Traits), "After I Made a Huge Mess with My Chemistry Set" (3rd Grade, Week 11: Force, Motion & Energy), "Orion Nebula" (3rd Grade, Week 14: Space), "Cancer" (3rd Grade, Week 35: Future Challenges); copyright ©2014 by Mary Lee Hahn. Used with permission of the author. All rights reserved.

**Avis Harley:** "Designing an Experiment: Will a Car Roll Faster Down a Steeper Slant?" (5th Grade, Week 6: Investigations); copyright ©2014 by Avis Harley. Used with permission of the author. All rights reserved.

**David L. Harrison:** "I Want to Know Why" (1st Grade, Week 35: Future Challenges), "My Robot" (2nd Grade, Week 29: Building Things); copyright ©2014 by David L. Harrison. Used with permission of the author. All rights reserved.

**Terry Webb Harshman:** "Queen of Night" (4th Grade, Week 15, Sun, Earth & Moon); copyright ©2014 by Terry Webb Harshman. Used with permission of the author. All rights reserved.

**Juanita Havill:** "Magic Show" (1st Grade, Week 19: Soil & Land), "Rings Not Letters" (2nd Grade, Week 24: Patterns), "Space Yacht" (2nd Grade, Week 36: Future Dreams), "The NEO Hunters" (4th Grade, Week 14: Space); copyright ©2014 by Juanita Havill. Used with permission of the author. All rights reserved.

**Esther Hershenhorn:** "What Am I?" (3rd Grade, Week 27: Video Technology); copyright ©2014 by Esther Hershenhorn. Used with permission of the author. All rights reserved.

**Sara Holbrook:** "Water Engineered" (2nd Grade, Week 34: Science Careers), "Friction" (4th Grade, Week 12: More Force, Motion & Energy), "Cool Food for Thought" (5th Grade, Week 21: Ecosystems); copyright ©2014 by Sara Holbrook. Used with permission of the author. All rights reserved.

**Mary Ann Hoberman:** "Trilobite" (3rd Grade, Week 19: Soil & Land); copyright ©2014 by Mary Ann Hoberman. Used with permission of the author. All rights reserved.

**Patricia Hubbell:** "Love Note to a Magnet" (1st Grade, Week 11: Force, Motion & Energy), "Roller Coaster Ride" (4th Grade, Week 11: Force, Motion & Energy); copyright ©2014 by Patricia Hubbell. Used with permission of the author. All rights reserved.

**Jacqueline Jules:** "Protecting My Friend" (3rd Grade, Week 25: Human Body), "Soda Machine Bite" (5th Grade, Week 28: Machines); copyright ©2014 by Jacqueline Jules. Used with permission of the author. All rights reserved.

**Bobbi Katz:** "The Rain Forest" (2nd Grade, Week 21: Ecosystems), "Lunar Eclipse" (3rd Grade, Week 15: Sun, Earth & Moon), "Considering Copernicus" (3rd Grade, Week 31: Famous Scientists); copyright ©2014 by Bobbi Katz. Used with permission of the author. All rights reserved.

**X.J. Kennedy:** "Metal Monster" (Kindergarten, Week 28: Machines), "Discovery" (1st Grade, Week 5: Predictions & Hypotheses); copyright ©2014 by X.J. Kennedy. Used with permission of the author. All rights reserved.

**Julie Larios:** "Did You Know?" (Kindergarten, Week 14: Space), "Rachel Carson" (Kindergarten, Week 31: Famous Scientists), "Testing My Magnet" (1st Grade, Week 7: Data), "My Experiment" (2nd Grade, Week 6: Investigations), "Albert Einstein" (4th Grade, Week 31: Famous Scientists); copyright ©2014 by Julie Larios. Used with permission of the author. All rights reserved.

**Irene Latham:** "Riddle for a Dry Day" (Kindergarten, Week 18: Forces of Nature), "Riddle for a Wet Day" (1st Grade, Week 18: Forces of Nature), "Science Fair" (4th Grade, Week 30: Science Fair); copyright ©2014 by Irene Latham. Used with permission of the author. All rights reserved.

**Renée M. LaTulippe:** "Celsius Thermometer" (1st Grade, Week 8: Tools of Science), "Pieces" (1st Grade, Week 27: Video Technology), "Driftwood Hut" (1st Grade, Week 29: Building Things), "Da Vinci Did It!" (1st Grade, Week 31: Famous Scientists), "Let's All Be Scientists!" (2nd Grade, Week 1: Scientific Practices), "Galileo Galilei" (3rd Grade, Week 32: More Famous Scientists), "Virtual Adventure" (4th Grade, Week 27: Video Technology); copyright ©2014 by Renée M. LaTulippe. Used with permission of the author. All rights reserved.

**Debbie Levy:** "Wiki Alert" (3rd Grade, Week 33, Computers); copyright ©2014 by Debbie Levy. Used with permission of the author. All rights reserved.

**J. Patrick Lewis:** "The 'Black Leonardo'" (1st Grade, Week 32: More Famous Scientists); copyright ©2014 by J. Patrick Lewis. Used with permission of Curtis Brown, Ltd. All rights reserved.

**George Ella Lyon:** "Hand-Me-Downs" (1st Grade, Week 4: Observations), "Meet Mr. Wizard" (3rd Grade, Week 6: Investigations); copyright ©2014 by George Ella Lyon. Used with permission of the author. All rights reserved.

**Guadalupe Garcia McCall:** "Sun-Kissed" / "Besado por el sol" (3rd Grade, Week 23: Cycles), "Oh Water, My Friend" / "Ay agua, mi amiga" (4th Grade, Week 16: The Water Cycle), "Cicada" / "Chicharra" (5th Grade, Week 23: Cycles); copyright ©2014 by Guadalupe Garcia McCall. Used with permission of the author. All rights reserved.

**Heidi Mordhorst:** "Cicada Magic" (4th Grade, Week 24: Patterns); copyright ©2014 by Heidi Mordhorst. Used with permission of the author. All rights reserved.

**Marilyn Nelson:** "A New Dinosaur" (5th Grade, Week 34: Science Careers); copyright ©2014 by Marilyn Nelson. Used with permission of the author. All rights reserved.

**Kenn Nesbitt:** "My Project for the Science Fair" (1st Grade, Week 30: Science Fair); copyright ©2014 by Kenn Nesbitt. Used with permission of the author. All rights reserved.

**Lesléa Newman:** "First Science Project" (1st Grade, Week 26: Kitchen Science), "The Leopard Cannot Change His Spots" (2nd Grade, Week 22: Adaptations & Traits); copyright ©2014 by Lesléa Newman. Used with permission of Curtis Brown, Ltd. All rights reserved.

**Eric Ode:** "When You Are a Scientist" (Kindergarten, Week 1: Scientific Practices), "Science Fair Day" (Kindergarten, Week 30: Science Fair), "Which Ones Will Float?" (3rd Grade, Week 1: Scientific Practices), "Science Fair Project" (3rd Grade, Week 30: Science Fair); copyright ©2014 by Eric Ode. Used with permission of the author. All rights reserved.

**Linda Sue Park:** "The Real Thing" (4th Grade, Week 36: Future Dreams), "Accidentally on Purpose" (5th Grade, Week 5: Predictions & Hypotheses); copyright ©2014 by Linda Sue Park. Used with permission of Curtis Brown, Ltd. All rights reserved.

**Ann Whitford Paul:** "Plates" (2nd Grade, Week 18: Forces of Nature), "Questions, Questions" (4th Grade, Week 3: Ask and Ask Again); copyright ©2014 by Ann Whitford Paul. Used with permission of the author. All rights reserved.

**Greg Pincus:** "Late Night Science Questions" (2nd Grade, Week 3: Ask and Ask Again); copyright ©2014 by Greg Pincus. Used with permission of the author. All rights reserved.

**Mary Quattlebaum:** "Breakfast Alchemy" (3rd Grade, Week 26, Kitchen Science); copyright ©2014 by Mary Quattlebaum. Used with permission of the author. All rights reserved.

**Heidi Bee Roemer:** "Liquids Can't Contain Themselves" (2nd Grade, Week 10: More Matter), "Questions That Matter" (3rd Grade, Week 9: Matter), "Welcome to the Science Lab" (5th Grade, Week 2: Lab Safety), "Going Bananas" (5th Grade, Week 7: Data), "Let Me Join You" (5th Grade, Week 25: Human Body), "Thirsty Measures" (5th Grade, Week 26: Kitchen Science); copyright ©2014 by Heidi Bee Roemer. Used with permission of the author. All rights reserved.

**Michael J. Rosen:** "Titan in Man's Seaweed" (5th Grade, Week 35: Future Challenges); copyright ©2014 by Michael J. Rosen. Used with permission of the author. All rights reserved.

**Deborah Ruddell:** "The Science Lab Pledge" (1st Grade, Week 2: Lab Safety); copyright ©2014 by Deborah Ruddell. Used with permission of Writers House LLC. All rights reserved.

**Laura Purdie Salas:** "Go Fly a Kite" (2nd Grade, Week 12: More Force, Motion & Energy), "Things to Do in Science Class" (3rd Grade, Week 2: Lab Safety), "What Can You Make from Carbon?" (4th Grade, Week 10: More Matter), "To the Eye" (4th Grade, Week 13: Light & Sound), "The Shadow Grows (and Shrinks, and Grows)" (5th Grade, Week 15: Sun, Earth & Moon), "The Great Pyramid of Giza" (5th Grade, Week 29: Building Things); copyright ©2014 by Laura Purdie Salas. Used with permission of the author. All rights reserved.

**Michael Salinger:** "Levers" (1st Grade, Week 28: Machines), "Gears" (2nd Grade, Week 28: Machines), "Sound Waves" (3rd Grade, Week 13: Light & Sound), "No Penguins Here" (5th Grade, Week 11: Force, Motion & Energy); copyright ©2014 by Michael Salinger. Used with permission of the author. All rights reserved.

**Glenn Schroeder:** "Frisbee" (1st Grade, Week 12: More Force, Motion & Energy); copyright ©2014 by Glenn Schroeder. Used with permission of the author. All rights reserved.

**Joyce Sidman:** "Gravity" (2nd Grade, Week 11: Force, Motion & Energy); copyright ©2014 by Joyce Sidman. Used with permission of the author. All rights reserved.

**Buffy Silverman:** "Think of an Atom" (5th Grade, Week 9: Matter); copyright ©2014 by Buffy Silverman. Used with permission of the author. All rights reserved.

**Marilyn Singer:** "Photosynthesis" (1st Grade, Week 23: Life Cycles), "Lift" (3rd Grade, Week 12: More Force, Motion & Energy); copyright ©2014 by Marilyn Singer. Used with permission of the author. All rights reserved.

**Ken Slesarik:** "My Rock" (Kindergarten, Week 19: Soil & Land); copyright ©2014 by Ken Slesarik. Used with permission of the author. All rights reserved.

**Eileen Spinelli:** "Thank You, Isaac Newton" (Kindergarten, Week 12: More Force, Matter & Energy), "What I Know about the Sun" (1st Grade, Week 15: Sun, Earth & Moon), "Imagine Small" (2nd Grade, Week 9: Matter); copyright ©2014 by Eileen Spinelli. Used with permission of the author. All rights reserved.

**Anastasia Suen:** "I Have a Question" (Kindergarten, Week 3, Ask and Ask Again), "Rain Gauge" (2nd Grade, Week 17: Weather & Climate); copyright ©2014 by Anastasia Suen. Used with permission of the author. All rights reserved.

**Susan Marie Swanson:** "Sound Waves at Breakfast" (2nd Grade, Week 13: Light & Sound), "Jane Goodall Begins a Speech" (2nd Grade, Week 31: Famous Scientists); copyright ©2014 by Susan Marie Swanson. Used with permission of the author. All rights reserved.

**Carmen Tafolla:** "Everyday Astronaut" / "Un astronauta común" (1st Grade, Week 36: Future Dreams), "My WristRobot Pack" (5th Grade, Week 36: Future Dreams); copyright ©2014 by Carmen Tafolla. Used with permission of the author. All rights reserved.

**Holly Thompson:** "Comet Hunter" (5th Grade, Week 14: Space); copyright ©2014 by Holly Thompson. Used with permission of the author. All rights reserved.

**Amy Ludwig VanDerwater:** "My Bean Plant" (Kindergarten, Week 7: Data), "Listen" (Kindergarten, Week 13: Light & Sound), "How to Be a Scientist" (1st Grade, Week 1: Scientific Practices), "Prism" (1st Grade, Week 13: Light & Sound), "Meter Stick" (2nd Grade, Week 8: Tools of Science); copyright ©2014 by Amy Ludwig VanDerwater. Used with permission of Curtis Brown, Ltd. All rights reserved.

**Lee Wardlaw:** "Science Project" (2nd Grade, Week 30: Science Fair); copyright ©2014 by Lee Wardlaw. Used with permission of Curtis Brown, Ltd. All rights reserved.

**Charles Waters:** "Mold" (2nd Grade, Week 26: Kitchen Science), "Froggy" (4th Grade, Week 6: Investigations), "Foundation (Don't Rush It!)" (4th Grade, Week 29: Building Things); copyright ©2014 by Charles Waters. Used with permission of the author. All rights reserved.

**April Halprin Wayland:** "Old Water" (Kindergarten, Week 16: The Water Cycle), "Can Our Eyes Fool Our Taste Buds?" (Kindergarten, Week 26: Kitchen Science), "I Like that Night Follows Day" (1st Grade, Week 24: Patterns), "Zapped!" (3rd Grade, Week 7: Data); copyright ©2014 by April Halprin Wayland. Used with permission of the author. All rights reserved.

**Carole Boston Weatherford:** "Harbor Wave at Hilo: Tsunami Survivor" (5th Grade, Week 18: Forces of Nature); copyright ©2014 by Carole Boston Weatherford. Used with permission of the author. All rights reserved.

**Steven Withrow:** "What Makes a Turbine Turn" (3rd Grade, Week 20, Natural Resources), "Moving to Atlantis City, 2112" (3rd Grade, Week 36, Future Dreams); copyright ©2014 by Steven Withrow. Used with permission of the author. All rights reserved.

**Allan Wolf:** "Step Outside. What Do You See?" (Kindergarten, Week 4: Observations); copyright ©2014 by Allan Wolf. Used with permission of the author. All rights reserved.

**Virginia Euwer Wolff:** "That Dish Thing" (5th Grade, Week 8: Tools of Science); copyright ©2014 by Virginia Euwer Wolff. Used with permission of Curtis Brown, Ltd. All rights reserved.

**Janet Wong:** "Sink or Float" (Kindergarten, Week 5: Predictions & Hypotheses), "Stopwatch" (Kindergarten, Week 8: Tools of Science), "Take Backs" (Kindergarten, Week 10: More Matter), "Push Power" (Kindergarten, Week 11: Force, Motion & Energy), "Auntie V's Hybrid Car" (Kindergarten, Week 20: Natural Resources), "Hello, Hello!" (Kindergarten, Week 27: Video Technology), "Tinker Time" (Kindergarten, Week 29: Building Things), "Dr. Lee" (Kindergarten, Week 34: Science Careers), "Future Dreams Idea #63" (Kindergarten, Week 36: Future Dreams), "Backwards" (1st Grade, Week 3: Ask and Ask Again), "Our Truck" (1st Grade, Week 9: Matter), "Sugar Water" (1st Grade, Week 10: More Matter), "Looking at the Sky Tonight" (1st Grade, Week 14: Space), "The Class Plant" (2nd Grade, Week 5: Predictions & Hypotheses), "Crazy Data Day" (2nd Grade, Week 7: Data), "Fossil Fuels" (2nd Grade, Week 20: Natural Resources), "Game Programmer" (2nd Grade, Week 27, Video Technology), "The Brink" (3rd Grade, Week 10: More Matter), "This Week's Weather" (3rd Grade, Week 17: Weather & Climate), "Computer Models" (4th Grade, Week 8: Tools of Science),

"Changes" (4th Grade, Week 9: Matter), "Hurricane Hideout" (4th Grade, Week 18: Forces of Nature), "Grafting" (4th Grade, Week 22: Adaptations & Traits), "Microwave Oven" (4th Grade, Week 28: Machines), "My Photo Experiment" (4th Grade, Week 33: Computers), "Paper Airplanes" (5th Grade, Week 3: Ask and Ask Again), "Frames Per Second (fps)" (5th Grade, Week 27: Video Technology), "Shen Kuo" (5th Grade, Week 31: Famous Scientists); copyright ©2014 by Janet S. Wong. Used with permission of the author. All rights reserved.

**Jane Yolen:** "Alligator with Fish" (Kindergarten, Week 21: Ecosystems), "The Lament of Lonesome George" (2nd Grade, Week 35: Future Challenges), "Tide Pool" (3rd Grade, Week 21: Ecosystems), "What Is a Foot?" (5th Grade, Week 22: Adaptations & Traits); copyright ©2014 by Jane Yolen. Used with permission of Curtis Brown, Ltd. All rights reserved.

## About the Poets

Biographical information, photos, and lists of some of the published titles of each of our contributing poets can be found at our website, PomeloBooks.com. Younger children might like to know, for instance, that April Halprin Wayland was once an aqua farmer and that Margarita Engle helps with wilderness search and rescue dog training programs.

Most poets have their own websites, too, where you can find contact info for them as well as news about their books and even links to their blogs. Some particularly useful poets' blogs are listed in the Poetry Resources section of this book.

If you identified "favorite poets" when reading the poems in this anthology, you might want to contact them about speaking at your school—either in person or via video chat—or about participating in a conference for teachers. Some poets enjoy large assemblies, some prefer small workshops, and some do both. Contact them and start a conversation!

## About This Book

# TGIF!

We hope that you are finding Fridays even more special now that you're taking a few minutes each week to share a science poem with your students on Poetry Friday. Spending time together with a thoughtful or interesting poem is a wonderful way to develop a classroom community, talk about our questions and experiences, and learn new words and science concepts. And if you've shared 36 poems (one per week throughout the school year), we hope it has also become a beloved tradition.

Have you been marking this book up with notes about your favorite poems and strategies for sharing poetry? If not, go back and do that now. Add whatever you remember about students' responses to individual poems. Jot down your thoughts about linking various poems with favorite science lessons or units. We want this book to become one of the most useful resources in your professional library—as well as one of your favorite books for sharing aloud.

Wouldn't it be wonderful if this Poetry Friday experience was part of every grade level for every student? What a culture of literacy and language love and curiosity that would create! Please help us spread the word about the power of poetry and ensure that every child has a chance to experience these fabulous five minutes every Friday!

## Acknowledgments

We have been so pleased with the response to our first two Poetry Friday anthologies (for K-5 and middle school) that we wanted to think BIG! What was the next logical step? To link poetry with other areas of the curriculum that we might not usually associate with poetry. Which area to consider first? We decided it had to be science! Upon reflection, this is not such an unusual pairing. For generations, poets have been observing nature, exploring the physical world, and asking questions about the universe.

It shouldn't be surprising that poetry has a lot to offer the sciences. In fact, astrophysicist Adam Frank revealed, "Poems and poetry are, for me, a deep form of *knowing*, just like science . . . each, in its way, is a way to understand the world." Poets and scientists both seek to observe, explain, and understand the world around them. So compiling an anthology of poems that were science-themed was a compelling invitation for all of the poets who participated in this publication, and we are so grateful for their wonderful, diverse contributions. Without the poets, we have nothing. Please get to know them and their books on their individual pages in the Meet the Poets section of our website, PomeloBooks.com.

But it was also very important to "get the science right." These are not only evocative and engaging poems, but we are also presenting them as a means for young readers to learn more about science. So we also took great care to "vet" the poems for science content. A gigantic *thank you* to our several science experts who reviewed our poems and Take 5! activities and gave us thoughtful feedback on how to improve and refine them. We especially want to recognize Shirley Smith Duke, Britt Bothe, Kathleen Hoke, and Mark VanDerwater. We also appreciate the close reading, additional research, and editorial support provided by the amazing Shirley Smith Duke (it bears repeating!).

We hope that this book helps teachers, librarians, and parents who want to encourage children to wonder about the natural world and also treasure the power of words and language. If you liked this book, please help spread the word—with your colleagues in your school and school district and community. If any of your fellow educators are not in the habit of sharing poetry or perhaps uncomfortable with the area of science, those are the people you need to talk to first!

A final note: our thanks to our readers and book buyers and book sharers for their support. Please visit our blogs and website and continue to spread the word about our books!

With sincerest appreciation,

*Sylvia and Janet*

PomeloBooks.com

# About Sylvia Vardell

**Sylvia M. Vardell** is Professor in the School of Library and Information Studies at Texas Woman's University and has taught graduate courses in children's and young adult literature at various universities since 1981. Vardell has published extensively, including five books on literature for children, as well as over 20 book chapters and 100 journal articles. Her current work focuses on poetry for children, including a regular blog, *PoetryforChildren,* since 2006. She is also the regular poetry columnist for ALA's *Book Links* magazine.

Vardell has served as a member or chair of several national award committees including the NCTE Award for Poetry, the NCTE Notables, the Cybils Poetry Award, the ALA Odyssey Award for audiobooks, the ALA Sibert Award for informational literature, and the NCTE Orbis Pictus Award for nonfiction, among others. She has conducted over 100 presentations at state, regional, national, and international conferences, and has received grants from the Young Adult Library Service Association (YALSA), Ezra Jack Keats Foundation, National Council of Teachers of English (NCTE), the Assembly on Literature for Young Adults Foundation, the Texas Library Association, and the National Endowment for the Humanities. She taught at the University of Zimbabwe in Africa as a Fulbright scholar and has served as a consultant to the Poetry Foundation.

## Other Professional Books by Sylvia Vardell

*Poetry Aloud Here 2: Sharing Poetry with Children* (2014)

*Children's Literature in Action: A Librarian's Guide* (2nd edition) (2014)

*The Poetry Friday Anthology for Middle School* with Janet Wong (2013)

*The Poetry Friday Anthology K-5* with Janet Wong (2012)

*The Poetry Teacher's Book of Lists* (2012)

*Poetry People: A Practical Guide to Children's Poets* (2007)

## About Janet Wong

**Janet S. Wong** is a graduate of Yale Law School and former lawyer who switched careers and became a children's poet. Her dramatic career change has been featured on *The Oprah Winfrey Show*, CNN's *Paula Zahn Show*, and *Radical Sabbatical*. She is the author of 30 books for children and teens on a wide variety of subjects, including writing and revision (*You Have to Write*), creative recycling (*The Dumpster Diver*), diversity and community (*Apple Pie 4th of July*), cheating on tests (*Me and Rolly Maloo*), and chess (*Alex and the Wednesday Chess Club*).

Wong has served as a member of several national committees including the NCTE Award for Poetry, the NCTE Commission on Literature, the Notable Books for a Global Society committee of the International Reading Association (IRA), the SCBWI Golden Kite committee (for picture books), and the PEN Center USA Literary Award committee (for children's literature). Wong is a frequent featured or keynote speaker at conferences and has worked with over 200,000 children at schools all over the world. Her recent focus is the exploration of digital opportunities for children's books; she encourages children not just to read e-books, but also to publish their own writing using affordable new technologies.

### Selected Poetry Books by Janet Wong

*Declaration of Interdependence: Poems for an Election Year* (2012)
*Once Upon a Tiger: New Beginnings for Endangered Animals* (2011)
*TWIST: Yoga Poems* (2007)
*Knock on Wood: Poems about Superstitions* (2003)
*Behind the Wheel: Poems about Driving* (1999)
*The Rainbow Hand: Poems about Mothers and Children* (1999)
*A Suitcase of Seaweed* (1996)
*Good Luck Gold* (1994)

# A Note About Our Student Editions

Be sure to connect your students and their families with our Student Editions for each grade so that they can revisit these poems on their own, at their own pace, as a fun supplement to their reading and learning activities.

Each grade-level Student Edition contains the 36 poems in this book's grade level, the "Poems for Everyone" (page 23 and page 265), and five Bonus Poems. The *Take 5!* teaching tips are **not** included in the Student Edition.

**Kindergarten Student Edition** (38 poems plus these below)

| | |
|---|---|
| Bonus Poem #1 | Life Story *by Linda Ashman* |
| Bonus Poem #2 | What Is a Fossil? *by Rebecca Kai Dotlich* |
| Bonus Poem #3 | I Thought I Built a Dog House *by Eric Ode* |
| Bonus Poem #4 | Can You Hear a Conch? *by Laura Purdie Salas* |
| Bonus Poem #5 | Butterfly Garden *by Jane Yolen* |

**First Grade Student Edition** (38 poems plus these below)

| | |
|---|---|
| Bonus Poem #1 | No Hurry *by Linda Ashman* |
| Bonus Poem #2 | Pollination/Polinización *by Margarita Engle* |
| Bonus Poem #3 | Lab Time *by Renée M. LaTulippe* |
| Bonus Poem #4 | Squiggles *by Michael Salinger* |
| Bonus Poem #5 | Seven Words about an Alligator *by Jane Yolen* |

**Second Grade Student Edition** (38 poems plus these below)

| | |
|---|---|
| Bonus Poem #1 | Power Blender and Aha! / Enchufa la batidora y ¡ya está! *by F. Isabel Campoy* |
| Bonus Poem #2 | Alexander Graham Bell *by Avis Harley* |
| Bonus Poem #3 | Disaster Riddle in a Hurry *by Irene Latham* |
| Bonus Poem #4 | Mrs. Sepuka's Classroom Pet *by Ken Slesarik* |
| Bonus Poem #5 | Snack *by Jane Yolen* |

**Third Grade Student Edition** (38 poems plus these below)

| | |
|---|---|
| Bonus Poem #1 | Happiness in the Desert *by Joy Acey* |
| Bonus Poem #2 | Seeking Proof *by Carole Gerber* |
| Bonus Poem #3 | Disaster Riddle Under Pressure *by Irene Latham* |
| Bonus Poem #4 | I Plan to Be an Astronaut *by Kenn Nesbitt* |
| Bonus Poem #5 | Sound Waves *by Amy Ludwig VanDerwater* |

**Fourth Grade Student Edition** (38 poems plus these below)

| | |
|---|---|
| Bonus Poem #1 | Mount St. Helens *by Carmen T. Bernier-Grand* |
| Bonus Poem #2 | Lunar Eclipse *by Avis Harley* |
| Bonus Poem #3 | Lorenzo Liszt, Non-Scientist *by Kenn Nesbitt* |
| Bonus Poem #4 | Seashells in the Sky *by Laura Purdie Salas* |
| Bonus Poem #5 | Story Rocks *by Susan Marie Swanson* |

**Fifth Grade Student Edition** (38 poems plus these below)

| | |
|---|---|
| Bonus Poem #1 | Stone, Sea, and Silence *by Jeannine Atkins* |
| Bonus Poem #2 | 3-D *by Betsy Franco* |
| Bonus Poem #3 | What Should I Call It?! *by J. Patrick Lewis* |
| Bonus Poem #4 | For the Science Fair … *by Ann Whitford Paul* |
| Bonus Poem #5 | MMO *by Anastasia Suen* |

Learn more about our Student Editions, e-books,
and other books in our series at
PomeloBooks.com.

# Praise for *The Poetry Friday Anthology* Series

**K-5 Poetry Edition**

"**Find a place for this book on your desk** since you'll be turning to it time and time again."

—Barbara Ward, *IRA's Reading Today*

"**This is a lot of resource and professional development** for $29.99!"

—Jeanette Larson, from *The ALSC Blog*

**Middle School Poetry Edition**

"**It's a *vade mecum* for the elementary teacher** and a word magnet for the K-5 child. Brava to the anthologists and the poets!"

—J. Patrick Lewis, Children's Poet Laureate

"The Common Core standards provided throughout the book **give teachers confidence that they are integrating key skills** as they share the poems. The book highlights and documents specific skills and techniques, such as rhyme, repetition, rhythm, and alliteration, as they are used one poem at a time."

—*U.S. Kids Magazine (Parents & Teachers)*

**Science Student Editions for K, 1, 2, 3, 4, 5**

"*Savvy teachers have learned they can trust Vardell and Wong.*"

—*IRA's Reading Today*

For more information about *The Poetry Friday Anthology* series, including Student Editions of this book, please visit PomeloBooks.com.